# Meatmaster Sheep
## Breed Establishment in South Africa

**Freddie Peters**

KeNa
KNOWLEDGE WORKS

Published by

Kejafa Knowledge Works

PO Box 7114
Krugersdorp North
1741
www.kejafa.co.za
kejafa@mweb.co.za

**Meatmaster Sheep Breed Establishment in South Africa**

First Edition 2011
First Print 2011

**ISBN: 978-0-9869983-1-7**
Cover design: Christina van Straaten
Page layout: Christina van Straaten
Language:  Franso-Mari Olivier
Printed and bound by:  Craft Print International Ltd, Singapore

# Meatmaster Sheep
## Breed Establishment in South Africa

by

**FW Peters** [1]

**August 2011**

1 Department of Zoology, University of Johannesburg, Private Bag 524, Auckland Park, 2006, South Africa

# Reviews

"This book is a comprehensive record of the development of the Meatmaster sheep as a recognized breed in South Africa and the scientific and other tools that were used by a group of dedicated breeders to develop the breed, without any direct government support.

The book starts with an introductory chapter that gives an extensive overview of the development of the breed. This is followed by five more chapters, each describing a specific aspect of the breed or its development. This includes a phenotypic description of the breed and a motivation for breed standards; the breed development program; genetic profile of the breed; characterization of the production potential; and lastly agro-economics of the breed. There are also a wide variety of photos that complement the contents of the book.

The book is well rounded off and attempts not to be controversial. It is an excellent register of the development of the breed.

Meatmaster Sheep Breed Development in South Africa should be on every mutton sheep farmer's shelf, as well as at agricultural colleges and related governmental institutions. It should also be in the libraries of universities. The excellent list of references, with correct scientific nomenclature, is applauded."
- **Prof. Michiel M Scholtz,** Specialist Researcher in Applied Animal Breeding at the ARC-Animal Production Institute and Extraordinary Professor in Animal Breeding at the University of the Free State

"My sincere congratulations to Freddie Peters for a work very well done. How you managed to gather all this information is admirable. As Meatmaster breeders we will benefit for years from all this information available to us. The history, the scientific information on blood typing and DNA, the comparison of breeds, can all help us answer so many questions asked to us daily.

Your book also focuses prominently on the fact that Meatmaster breeding is all about breeding a more profitable sheep for the commercial industry. It is not just another beautiful sheep to show with, that will be popular for a while because it is something new but then slowly lose popularity as numbers increase.

Reading and understanding this book and applying with wisdom the contents will convince diligent sheep breeders that Meatmasters are here to stay and serve meat sheep farmers in great numbers with rams that will keep them economically viable.

Most importantly this book will help future breeders to return to the principles of breeding and selection that laid the foundation for the present greatness of the Meatmaster should they ever go astray. Once again thank you and congratulations Freddie. May the Meatmasters prosper." - **Clynton Collett,** Council Member and first President of the Meatmaster Sheep Breeders' Society of South Africa

"It has become increasingly apparent that the development and establishment of the officially recognized Meatmaster sheep breed should be documented comprehensively.

The work put in and passion of the author regarding the development and establishment of the Meatmaster sheep breed over many years is commendable. His enthusiasm for the breed is contagious and his efforts are extensively reported in a highly practical and accessible manner. It is also a tribute to the many collaborators who shared the same dream, most of whom are acknowledged and shown on its pages. Another dimension is provided by many interesting historical details and an abundance of colorful illustrations. This is a comprehensive and unique documentation of the Meatmasters and a wider range of readers would benefit from this informative narrative, than the obvious intended target of those interested in the breed, current and prospective breeders, Breeding Societies and students. This book will thus serve as a worthwhile reference for future generations and contribute to the body of knowledge of the Meatmaster breed.

This book serves as a testimony to the author's passion and concern about the existence and improvement of the Meatmasters." - **Prof Japie (B) van Wyk,** Department of Animal, Wildlife and Grassland Sciences, University of the Free State

"In our current agricultural environment, the need for an easy care, economically viable sheep breed capable of efficient production and reproduction from natural veld conditions is indisputable. The Meatmaster was developed by some farmers to fulfill this need. In this book, the development of the Meatmaster breed is documented in an interesting and informative way. The author's commitment to and passion for the Meatmaster sheep speak from every chapter of this book.

To have such a comprehensive work available, which include the entire spectrum from breed development and history, through to breed standards, phenotypic standards based on measured performance as well as a genetic analysis of the breed, is a first for a South African sheep breed. The fact that the Meatmaster is the youngest locally developed breed makes this much more remarkable and the author and all the involved breeders could be very proud of this achievement.

Not only Meatmaster and other mutton sheep producers will benefit from this book, but it will be very informative for both agricultural college and university students. As some of the chapters in this book were also included in a research publication, the book will also be of scientific value to researchers. The extensive reference list furthermore contributes to the scientific value of this publication." - **Dr Gretha (M A) Snyman**, Specialist Agricultural Scientist, Grootfontein Agricultural Development Institute, Middelburg (EC)

# Dedication

This book is dedicated to Clynton Collett, a leader who leads by example, a friend who supports with inspiration, a humble servant of the Lord, a man dedicated to meet the commitments of his cause, to our Meatmaster breeders the kingpin and the soul of the Society; and to the memory of Christine du Toit who was a humble pillar of strength and support who always paved the way ahead without expecting anything in return.

*Clynton Collett, founder member and first President of the Meatmaster Sheep Breeders' Society of South Africa.*

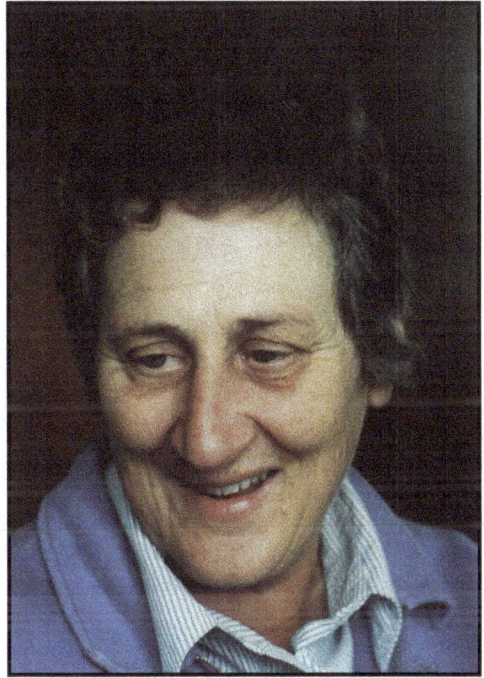

*The late Christine du Toit, founder member and first Vice-President of the Meatmaster Sheep Breeders' Society of South Africa.*

# Acknowledgments

Due to the comprehensive nature of this book many contributions were made by family, friends, colleagues and institutions which I would gladly like to acknowledge:

- Dr MA Snyman from the Grootfontein Agricultural Development Institute (GADI) and Prof Dr MM Scholtz, extraordinary professor at the University of the Free State and specialist researcher with the ARC at Irene, for guidance in animal science, assistance with statistical analysis of Meatmaster performance data and interpretation of results.

- Mr Clynton Collett for the wealth of information and documentation supplied and for his tireless efforts in the establishment, promotion and improvement of the Meatmaster sheep breed.

- Dr Walti Vermeulen and the late Ms Christine du Toit for personal communications, valuable information supplied and assistance with the collection of Meatmaster sheep blood samples from their respective flocks of Meatmaster sheep in the Northern Cape areas of Prieska and Hopetown.

- The Agricultural Research Council (ARC) Animal Genetics Laboratory at Irene for the support and assistance with the DNA analysis of sheep blood samples. Mr Hans van Zyl for the dedicated assistance to apply the Lyx program to the writing of the initial documents and Ms Pranisha Soma for assistance with the DNA profiles of the 256 sheep blood samples analysed.

- The contributions of A Kotze, FH van der Bank, P Soma and JP Grobler in a peer reviewed article titled *"Genetic profile of the locally developed Meatmaster sheep breed in South Africa based on micro-satellite analysis"* which was published in Small Ruminant Research 90:101-108 (2010), cited as (FW Peters et al., 2010) and which is incorporated into chapter 4 of this book.

- Prof Dr M Wink from the Heidelberg University in Germany, Prof JB van Wyk from the University of the Free State and Dr E van Marle-Köster from the University of Pretoria for valuable comments, suggestions and inputs by which the book was improved.

- Prof Herman van der Bank from the University of Johannesburg for his assistance and Prof Antoinette Kotze associated with the University of the Free State and the National Zoological Gardens, for her efforts to bring the Meatmaster project to the attention of National as well as International interest groups.

- Prof Paul Grobler from the University of the Free State for the valuable contributions and assistance with the statistical analysis of the Meatmaster sheep DNA data and for his efforts as corresponding author of the research article on the Meatmaster sheep genetic profile.

- The Gellap-Ost agricultural research station at Keetmanshoop in Namibia for the opportunity to compare the DNA data of the Gellapper sheep population to the Meatmaster DNA database and for the use of the recorded production data on the Gellapper, Dorper and Damara parent breeds.

- Ms Lysette Pretorius-Peters for the hours spent in assistance with the preparation of PowerPoint presentations and the editing of photographs.

- Ms Naomi Strydom and Ms Christa Coetzee from the University of Johannesburg for administrative support, typing, editing and formatting of documents used in the book.

- Mr Richard Devey from Statkon at the University of Johannesburg for assistance with statistical analysis of recorded data.

- Mr Mogobothi Linda Zozo, Mr Tsokolo Leonard Makoaqa and the late Mr Johannes Stokile Mokoena for the guarding of the Meinfred Meatmaster flock of sheep and for assistance with the collection of blood samples, the marking, recording of phenotypic data and general care of Meatmasters bred during 2005 to 2009.

- Mr Mario du Preez for the use of photographs to provide a visual description of Meatmaster sheep and breed standards.

- Ms Martie Wallace from the University of Johannesburg for typing and preparation of documents and appendices used in the book.

# Summary

## Meatmaster Sheep Breed Establishment in South Africa

Based on the vision that by combining the characteristics of durability and fertility exhibited by the Damara sheep of Southern Africa and the much sought after conformation of the Ile de France and Dorper types of sheep, a new composite mutton sheep breed could be created which would meet the requirements for an easy care, economically viable and widely adaptable mutton sheep, a number of breed creation programs were implemented. The progeny thus produced was evaluated over a period of time with regard to fecundity, productivity, carcass quality and market acceptability. This resulted in the establishment of distinct populations of mutton sheep originally referred to as Damara crossbred sheep (Hofmeyer, 2001) and Bosrander sheep (Cilliers, 2000) because they adapted extremely well in the mountainous areas of the Eastern Cape and the bushy and mountainous areas of the Suikerbosrand near Meyerton in the Gauteng Province of South Africa.

The Meyerton population of Bosrander sheep ultimately became known as the Meinfred Meatmaster sheep stud (Muller, 2007). Following the publication of an article in the Landbouweekblad of 29 September 2000, it became evident that there were other sheep breeders, notably in the Eastern and Northern Cape Provinces of South Africa also attempting to establish new sheep breeds from the initial crossing of Damara sheep with inter alia Dorper, Van Rooy and Mutton Merino Sheep. This brought about the realisation that in order to establish a formally recognised new sheep breed, it was necessary to acquire sufficient consensus amongst prospective breeders to synchronise their breeding goals in order to breed what then became known as the Meatmaster sheep breed.

The Meatmaster Sheep Breeders' Society of South Africa was formally established on 4 February 2005 at Brandford in the Free State. The first executive committee, elected on 16 February 2005 was: CR Collett (President), Ms CM du Toit (Vice President), FW Peters (Secretary) and JP du Plessis.

The new Meatmaster breed was presented by FW Peters to the international delegates of the 6th Global Conference on the Conservation of Farm Animal Genetic Resources during October 2005 and to the South African Society of Animal Science Symposium held at Upington during July 2006. CR Collett presented the Meatmaster breed to the International Clean Skin Symposium at Adelaide in Australia during February 2010.

In order to prove that the Meatmaster is a genetically recognizable group, blood samples of four Meatmaster sheep populations were taken by FW Peters during 2005 and 2006 and prepared for DNA analysis at the ARC Animal Genetics Laboratory at Irene in order to establish genetic profiles of the new Meatmaster sheep breed. Statistical analysis of the DNA data and comparison of the data with the DNA profiles of parent breeds was done at the University of the Free State during 2007 and 2008. The results obtained proved that the Meatmaster was a genetically recognizable group.

The first opportunity for the application of the Meatmaster DNA database presented itself early during 2009 when a request was received by FW Peters from the Department of Agriculture of Namibia to compare the DNA data from their Gellapper sheep, bred during research trials at the Gellap-Ost research station near Keetmanshoop, to the Meatmaster DNA data and to advise them on the way forward.

A research article *"Genetic profile of the locally developed Meatmaster sheep breed in South Africa based on microsatellite analysis"*, cited as FW Peters et al., was published in Small Ruminant Research in May 2010.

The work done by the first Board of the Meatmaster Sheep Breeders' Society of South Africa made a decisive contribution to the establishment of the Meatmaster sheep breed, which was formally recognised and duly Gazetted by the Registrar of Animal Improvement in Government Gazette No. 29898 dated 25 May 2007. A second proclamation by the National Department of Agriculture in South Africa was gazetted by the Registrar in Government Gazette No. 32601, dated 2 October

2009, in which the Meatmaster sheep breed was reclassified as a recognised locally adapted and regularly introduced sheep breed alongside well known sheep breeds such as the Suffolk, Ile de France and Merino land sheep (Appendix B).

The phenotypic characterization of productive traits of the new Meatmaster breed was done in collaboration with Dr MA Snyman and Dr JJ Olivier from the Grootfontein Agricultural Development Institute (GADI) and Dr MM Scholtz from the ARC at Irene in order for the breed to be formally accepted in the environment of animal science. Reproductive data recorded on the farm La Rochelle in the Eastern Cape between 1999 and 2009 were obtained from the National Small Stock Improvement Scheme and analysed with the SAS program in collaboration with the Grootfontein Agricultural Development Institute (GADI). Breeding values obtained from the INTERGIS were used to obtain genetic trends.

Productive traits recorded for the Meatmaster sheep at three different localities (Highveld, Eastern Cape and Keetmanshoop) were compared to the phenotypic values of the parent breeds and were statistically analysed.

Membership of the Meatmaster Breeders' Society established by three members during February 2005, gradually increased to sixteen members by the end of 2006, but by the end of December 2009 the membership list of Meatmaster Breeders (Appendix A) increased to sixty-five registered members, thereby becoming the fourth largest sheep breeders' society amongst the ten mutton and dual purpose sheep breeds in South Africa.

An agro-economical evaluation of the Meatmaster breed was done by application of the SM2000 program developed at GADI by Herselman (2002), for the three different localities where production data were recorded in order to prove the profitability of the new Meatmaster breed.

## Key Terms

Sheep breeds; Ovine diversity; Meatmaster; Genetic structure; Breed creation; Phenotypic characterization; Agro-economics

# Table of Contents

# CHAPTER 5 CHARACTERIZATION OF THE PRODUCTION POTENTIAL OF THE MEATMASTER SHEEP

# APPENDICES .......................................................................................................... **98**

# List of Tables

## CHAPTER 6

# List of Figures

## CHAPTER 3

## CHAPTER 4

## CHAPTER 5

## CHAPTER 6

# Chapter 1 Introduction

## 1.1 General

The International Livestock Research Institute (ILRI) pointed out that a livestock revolution has to take place to meet the demands for food of animal origin. *"The productivity of farm animals must increase to avoid overgrazing and subsequent degradation of natural resources"*. The Afrino was the last composite white woolled sheep breed developed in South Africa by a project initiated at the Carnarvon Experimental Station in the Northwestern Karoo. The Afrino is a three-breed composite consisting of 25% Merino, 25% Ronderib Afrikaner and 50% South African Mutton Merino (DAGRIS, 2005). Remarkable genetic improvement was made since the establishment of the Afrino Sheep Breeders' Society in 1980.

By utilizing performance testing information, it was possible to genetically increase body weight and lifetime total weight of lamb weaned, while simultaneously decreasing fibre diameter in the Carnarvon Afrino flock (Snyman et al., 1997; Snyman et al., 1998). Animal scientists furthermore attempt to achieve breed complementarity or sire x dam complementarity by selecting rams on growth and fleece traits and ewes on reproductive traits (Herselman et al., 1998).

Since the late 1980's a number of sheep farmers realized that there was a national as well as an international need for an adaptable, economically viable and marketable mutton sheep which can produce on the veld with minimum care and low input costs. *"The economic problem with fat-tailed sheep varieties has always been a poor carcass dominated by hind quarter fat localisation although they are known to be well adapted to harsh environmental conditions"* (Burgess, 2006).

The Meatmaster sheep breed will be developed to fulfill this need and to provide an acceptable carcass while adapting extremely well in harsh environmental conditions. Breed standards are based on agro-economic principles as opposed to breed standards developed for the show ring, which is still held in high esteem by many sheep stud breeders. Strong interest from neighbouring countries and international enquiries regarding the new Meatmaster sheep already indicate the viability for the export of much sought after genetic material.

The Meatmaster sheep breed will be one of only a few locally developed sheep breeds in the world that was developed by private individuals without state support. The project was also not funded by Agricultural Research Institutes. The development and establishment of the new Meatmaster sheep breed is the result of long-term dedication and financial sacrifice of a small number of individual sheep breeders who strived towards the attainment of the initial vision of an economically viable and generally acceptable easy care sheep breed.

The lack of involvement of experienced animal breeding scientists in the development of the new Meatmaster sheep breed is a matter of concern to the Meatmaster Sheep Breeders' Society. The documentation and information provided with regard to the Meatmaster sheep breed, although in some instances not yet scientifically proven, may stimulate the interest of animal breeding scientists to conduct further research and to contribute to the development of this new and unique sheep breed.

The initial objective of the Meatmaster Breeders' Society was essentially a practical one, and that was to contribute to the establishment and official recognition of the Meatmaster sheep breed.

### 1.1.1 Mutton sheep improvement in South Africa

Attempts to improve the weight and carcass quality of South African mutton sheep were made since the 1920's. Seymore (1937) mentioned that the comparison of 6 month old lambs of Persian, Merino and Persian-Suffolk cross lambs at the Grootfontein Agricultural College in 1929, yielded promising results. The crossings with exotic British mutton sheep also improved the carcass quality. Similar results were obtained with Persian x Dorset Horn and Persian x Southdown lambs.

Dr. QP Campbell who presented his DSc (Agric) Thesis (Campbell, 1974) on *"A study of breeding problems in Dorper sheep"* included a chapter as a review of mutton production in South Africa. In the Journal, Research Highlights - Animal Production, Campbell (1986) wrote an article titled *"The Dorper - a success story of stock improvement"*, pointing out that the development of the Dorper sheep breed was a good example of co-operation between the Meat Board, the Department of Agriculture and a number of sheep farmers. A surplus of mutton and lamb was produced in the late thirties before the Second World War. In 1935 there were 30 100 000 woolled sheep and 5 600 000 non-woolled sheep in South Africa. A surplus of mutton and lamb was supplied which was exported to the Smithfield market in London. Unfortunately the South African breeds such as the Blackhead Persian and Ronderib Blinkhaar Afrikaner lambs as well as the lean Merino lambs could not compete with the Canterbury lamb carcasses from New Zealand. The Meat Board consequently imported rams of the Dorset Horn, Suffolk, Texel, Welsh Mountain, South Down and Ryeland sheep breeds. These rams were made available to sheep breeders in the Karoo and the North-Western Cape to crossbreed them especially with Merino, Blackhead Persian and Ronderib Afrikaner ewes so as to procure suitable slaughter lambs. The crossbreeding was controlled by Messrs Dawid Engela and Harry Bonsma from the Grootfontein College of Agriculture.

Campbell (1988) furthermore pointed out in an article in the Dorpernuus that he informed Dorper breeders in a letter dated 1 June 1977 that he did in fact statistically prove that the modern type of Dorper was too dry, too woolly and conformed too much to the exotic Dorset Horn breed, resulting in the fact that the animals are more susceptible to lung, liver and hart diseases and also to skin irritation caused by external parasites. This points to genetic drift in the composite Dorper breed. Such genetic drift will inevitably have a great influence on the agro-economic aspects of any composite breed and may ultimately fail to meet the aim for which the breed was originally developed, simply because it is genetically not the same breed any more. According to Bourdon (2000) this problem can be *"undone"* by reconstituting the composite from time to time by adding to the composite population new first-generation composites whose purebred parents are unrelated to the purebreds that formed the foundation for the original composite population.

In the development of composite breeds it will become imperative in the future to have a clear genetic reference profile that could be used as a guideline to conserve its agro-economic value. *"Composite animals have at least two breeds in their background and often more. What distinguishes them from typical crossbreds is not their genetic makeup per se, but rather the way in which they are used. Composites are expected to be bred to their own kind, retaining a level of hybrid vigor normally associated with traditional crossbreeding systems, but without crossbreeding."* (Bourdon, 2000). In the development of the Meatmaster sheep breed a genetic reference profile for the Meatmaster breed is provided. Future Meatmaster generations can be measured against such a genetic profile in order to prevent undesirable genetic drift.

In the development of the Meatmaster sheep breed agro-economic factors were used to determine breed standards. The agro-economic value of the breed will be optimised by this approach. Much of the growing popularity of the Meatmaster sheep breed is vested in its perceived superior agro-economic value as a meat sheep breed.

### 1.1.2 Engineering Africa's ultimate veld sheep

*Fig 1.1   Beautiful examples of a Damara ram and ewe from the stud of Dawie du Toit at Prieska.*

In search of genetic candidates for the enginee-ring of Africa's ultimate veld sheep, indigenous sheep breeds such as the Namaqua Afrikaner, Blinkhaar Ronderib Afrikaner, Blackhead and Speckled Persians, and the Damara were evalu-ated. The Damara was found to be the breed that offered the most potential in terms of contribu-ting to the desired economic characteristics (Bur-gess, 2006). *"Selection of the breeds and the propor-tions of those breeds going into a composite is the critical step in composite breed formation and may well determine whether a composite breed succeed or fail. If the composite is put together in such a way that it exhibits close to optimum performance in the economically important traits when it is first formed, then any genetic change following breed formation can be considered fine-tuning."* (Bourdon, 2000).

The agro-economic factors that form the basis of the development of the Meatmaster sheep breed will ensure the future and popularity of this new sheep breed. The attractive character and protec-tive behavioural patterns of the Meatmaster com-bined with non selective grazing habits will make it a winner.

The Meatmaster sheep breed will be classified as a hair sheep breed. According to the Oklahoma State University data base of hair sheep breeds (2007/05/10) [http://www.ansi.okstate.edubreeds /sheep/hair.htm||Oklahoma state university] the following breeds are among hair sheep breeds listed:

- Barbados Blackbelly
- Blackhead Persian
- Damara
- Dorper
- Katahdin
- Red Maasai
- Wiltshire Horn

The author is of the opinion that from these hair sheep breeds, four could be considered to make a possible future contribution to the Meatmaster sheep if increased genetic diversity should be de-sired. The Barbados Blackbelly, a pure hair sheep breed with an established fecundity of 1.5 to 2.3, could be considered to increase fecundity and the Katahdin which is already a composite breed, de-veloped from the St. Croix and the Wiltshire horn sheep, could enhance conformation and thinner tails. The Wiltshire Horn is already incorporated in

the Meatmaster sheep breed as discussed under the heading *"Breed creation programs"*. The Red Maasai sheep which are indigenous to the sub-humid tropics in Kenya has been found to exhibit genetic resistance to gastro-intestinal nematode parasites (Baker et al., 1999) and could be consi-dered to ad to the future genetic make up of Meatmasters in sub tropical areas.

### 1.1.3 Genetic characterization of the Meat-master breed

Bududram (2004) pointed out that *"the Food and Agricultural Organization (FAO) of the United Na-tions proposed a global program for the manage-ment of genetic resources using molecular meth-odology for breed characterization. The study of the genetic diversity in a breeding population can help to determine the similarity of the genetic mate-rial carried by populations and the genetic variation they possess."* The Meatmaster breed was geneti-cally characterized as a unique sheep breed and populations of Meatmaster sheep will classify as Meatmasters within the genetic parameters of the breed.

#### 1.1.3.1 *Blood typing and protein polymor-phisms*

Blood typing and protein polymorphisms were often used during the 1960's but revealed a limi-ted number of loci and alleles at a locus (Nei, 1987). This method is rapid, affordable and reli-able, but requires fresh blood samples.

#### 1.1.3.2 *DNA micro-satellite markers*

The first study published using micro-satellite DNA markers for the genetic characterization of Southern African sheep breeds was conducted by Buduram (2004). Thirteen indigenous and locally developed sheep breeds were genetically charac-terised and included the Damara, Karakul, Pedi, Blinkhaar Ronderib Afrikaner, Van Rooy, Black-head Persian, Blackhead Speckled Persian, Red-head Persian, Redhead Speckled Persian, Zulu, Swazi, Namaqua Afrikaner and Dorper breeds. Seven merino wool types were also genetically characterised. These sheep breeds were the Dormer, SA Merino, SA Mutton Merino, Merino Landsheep, Letelle, Dohne Merino and the Afrino. DNA markers have a potential application over a relatively broad field in animal breeding and genetics. *"In breeding, quantitive trait loci (QTL) will for example be applied in the genotypic selection of superior animals. DNA technology is now well developed and provides many opportunities for genetic improvement in livestock in combination with the established quantitative methodologies."* (van Marle-Köster et al., 2003).

DNA micro-satellite markers were used for the genetic characterization of the Meatmaster sheep breed. This provides a unique opportunity to establish a genetic profile of the breed against which possible future genetic drift can be measured. The economic edge that the new Meatmaster sheep may have will be dependant on the performance in the economically important traits that stems from its genetic make-up.

Figure 1.2   Typical Meatmaster sheep

## 1.2   Brief History of the Meatmaster sheep breed in South Africa

Initially a number of breed creation programs were implemented by sheep breeders at different localities in South Africa, all of which included components of the Damara sheep breed in their new composite populations. This resulted in the establishment of distinct populations of mutton sheep originally referred to as Damara crossbred sheep (Hofmeyer, 2001) and Bosrander sheep (Cilliers, 2000) because they adapted extremely well in the mountainous areas of the Eastern Cape and the bushy and mountainous areas of the Suikerbosrand near Meyerton in the Gauteng Province of South Africa. The Meyerton population of Bosrander sheep bred by FW Peters since the early 1990's ultimately became known as the Meinfred Meatmaster sheep stud (Muller, 2007).

Figure 1.3   Freddie Peters with initial Highveld Meatmasters (2004)

This initial composite Meatmaster population consisted of Damara and Ile de France components and the flock numbered 460. Following the publication of an article in the Landbouweekblad of 29 September 2000, it became evident that there were other sheep breeders, notably in the Eastern and Northern Cape Provinces of South Africa also attempting to establish new sheep breeds from the initial crossing of Damara sheep with inter alia Dorper, Van Rooy and Mutton Merino Sheep (Hofmeyer, 2001). Also during the 1990's CR Collett in the Eastern Cape established a Meatmaster composite from Damara and Dorper parent breeds with a flock total of 3000 sheep.

Figure 1.4   Clynton Collett and Jean du Plessis with Eastern Cape Meatmasters at La Rochelle (Feb 2005).

Dr PvdW Vermeulen and family in the Hopetown district likewise established a large Meatmaster population with Damara, Dorper and Van Rooy components in a flock of 5000 sheep. Ms CM du Toit in the Prieska district established a Meatmaster flock of 1060 from Damara and Mutton Merino parent breeds which were subsequently mated to

Meatmasters from Damara x Dorper origin. Dr JJ Steyn and R Liebenberg bred a flock of about 500 Meatmasters from Damara and Dorper parents in the Bloemfontein area and subsequently introduced some Witshire Horn genetics to the flock. J Morrison, F Steyn and JW Swanepoel also became stakeholders in this Meatmaster flock. JAS Zwiegers bred 500 Meatmasters from Damara and Dorper parents in the Hopetown district. Other initial Meatmaster breeders in the Hopetown area were JP du Plessis and R Wiid.

*Figure 1.5   Jean du Plessis and Dr Walti Vermeulen at a Meatmaster training course*

DH Visser in the Kenhardt district of the Northern Cape established a flock of 250 Meatmasters from Damara and Dorper parents. D and J Steenkamp bred a Meatmaster flock near Wiliston since 1992 and in the Loeriesfontein district J and H Kearney and P Spangenberg were also involved with early Meatmaster breeding from Damara and Dorper parents.

This brought about the realisation that in order to establish a formally recognised new sheep breed, it was necessary to acquire sufficient consensus amongst prospective breeders to synchronise their breeding goals in order to breed what then became known as the Meatmaster sheep breed. CR Collett from the Eastern Cape became the kingpin in the coordination of Meatmaster sheep breeders. Meatmaster breed standards were subsequently defined and Meatmaster breeders were finally organised to form the Meatmaster Sheep Breeders' Society of South Africa.

## 1.2.1   Organised action to establish the new Meatmaster sheep breed

On 23 May 2000 a meeting of a number of sheep breeders, organized by CR Collett, was held at La Rochelle near Venterstad with the view to establish a new Breeders' Society for the Damara Crossbred Sheep. An interim Breeders' Committee was simultaneously elected. It was also decided that the breed name to be used will be the Meatmaster.

The Damara Breeders' Society of South Africa signed a cooperation agreement with the Meatmaster Committee on 30 April 2002 at Bloemfontein and the Damara and Meatmaster Sheep Breeders' Society of South Africa was officially founded on 2 October 2002 at Bloemfontein.

The Damara and Meatmaster Sheep Breeders' Society was officially dissolved on 4 February 2005 by a Special General Meeting and it was decided that two independent Breeders' Societies would be formed. Meatmaster breeders immediately proceeded to approve an interim constitution and to establish the Meatmaster Sheep Breeders' Society of South Africa. The founder members present were CR Collett, CM du Toit and FW Peters.

The first production sale of 600 Meatmaster ewes and 70 rams was held at La Rochelle, in the Venterstad district on 15 February 2005 and on 16 February 2005 the first General meeting of the Meatmaster Sheep Breeders' Society of South Africa was held where the first executive committee or Meatmaster council was elected. The inaugural council of the Meatmaster Breeders' Society consisted of four members, CR Collett as President, CM du Toit as Vice-President, FW Peters as secretary and JP du Plessis as additional member.

It was then decided to organise a two-day workshop where Meatmaster breeders could discuss the way forward and re-evaluate breed standards. The workshop was held from 31 May to 1 June 2006 at La Rochelle in the Eastern Cape.

Figure 1.6   Breeders evaluating Meatmaster breed standards – 31 May 2006
**Front (L-R)** Hester du Toit, Dr Buks (JJ) Olivier. **Middle:** Graham Howell, Freddie Peters, Clynton Collett, Christine du Toit, Marcel van der Merwe, Dr Johan Steyn, Johnny Morison **Back:** Unknown, Riaan Liebenberg, Mario du Preez

During April 2006 FW Peters submitted a 60 page application and motivation to the Registrar of Animal Improvement at the National Department of Agriculture in Pretoria for the proclamation of the Meatmaster sheep breed in terms of Section 2 (2) of the Animal Improvement Act (Act No. 62 of 1998) as well as an application for the formal registration of the Meatmaster Sheep Breeders' Society of South Africa in terms of Section 8 (2) of the Act. (Appendix B)

### 1.2.1.1   The first national sale of Meatmaster sheep and second AGM

The first National Sale of Meatmaster sheep was offered at Bloemfontein on 9 August 2006 and was preceded by the second Annual General Meeting (AGM) of the Meatmaster Sheep Breeders' Society of South Africa. The efforts of FW Peters to have the Meatmaster breed formally recognised and registered by the National Department of Agriculture was noted as well as his efforts to present the characteristics of the new breed at National and International Conferences. The executive committee for the Breeders' Society was re-elected as previously established. 260 Meatmaster ewes and 40 rams were offered for sale by six Meatmaster breeders.

### 1.2.1.2   Expansion of website information and links to Meatmaster information

Following the second AGM a new website was created for the Meatmaster sheep breed by

Studbreeder.com. Collett and Peters were listed as premium members with links to additional Meatmaster websites (Collett, 2005; Peters, 2007). Ms CM du Toit, listed as a sheep breeder with SA Studbook also established a link to the Geelbeksdam Meatmaster website (Du Toit, 2005) and the Meinfred Meatmaster stud was listed under the SA Studbook sheep breeders' list (SA Studbook, 2007).

### 1.2.1.3   International interest in Meatmasters and first exports of Meatmaster sheep

Following visits to Australia and Brazil during 2006, CR Collett reported that there was a national as well as an international need for an improved easy care hair sheep breed and that the Meatmaster could be the sheep to meet that need. Meatmasters are already being bred in Australia from Dorper and Damara sheep imported from South Africa. Sheep breeders in Canada and the USA are also interested in breeding Meatmaster sheep.

During the latter half of 2006 a number of Meatmaster rams were exported to Namibia from the Kenhardt district in the Northern Cape and during January 2007 two Meatmaster rams were exported to Botswana. Early in 2008 permission was obtained from the National Department of Agriculture in South Africa to export a number of performance recorded Meatmaster sheep to Namibia.

During July 2009 FW Peters visited the Agricultural Research station at Gellap-Ost in the Keetmanshoop district of Namibia following a request from the Department of Agriculture of Namibia to advise them on their sheep breeding research project involving the Gellapper sheep which was also bred from Damara and Dorper sheep but selected on the basis of different criteria. Delegates were sent from Windhoek, Mariental and Keetmanshoop to acquire information on the Meatmaster sheep breed and the prospects to incorporate the Gellapper flock of sheep into the Meatmaster breed was discussed, following the comparison of the DNA data of the Gellapper flock to the DNA database of the Meatmaster breed in South Africa.

### 1.2.1.4  *Fast growing popularity of the Meatmaster sheep breed (2005-2011)*

Prices obtained at Meatmaster sales provide a good indication of the general market acceptability of Meatmaster sheep. A Meatmaster ram CRC 4403 was sold for R32 000 at a production sale held by CR Collett in February 2008 indicating the growing popularity and acceptability of the new Meatmaster sheep breed. Several Meatmaster rams were subsequently sold for more than R20 000 and an increasing number of rams are sold at R10 000 to R20 000 at national Meatmaster sales. A number of Meatmaster ewes were sold for more than R10 000 and many reached between R2 000 and R3 000 at national sales. There is a general trend of growing interest in the new Meatmaster breed nationally and internationally.

*Figure 1.7   Meatmaster ram CRC 4403 sold for R32 000 in 2008*

*Figure 1.8  Meatmaster ram CRC 7405 was sold for R20 000 to a buyer in Namibia*

*Figure 1.9  Meatmaster ewe CRC 3259 with triplets sold for R13 500 in 2010*

Statistics from the production sale held by CR Collett on 9 Feb 2010 in the Eastern Cape provided further proof of the general market acceptability of the Meatmaster breed. 260 Meatmaster sheep sold achieved the under mentioned average prices. Buyers from all over South Africa, Namibia and Botswana attended the sale.

- 8 Proven sires averaged - R16 312
- 42 Stud rams averaged - R6 157
- 21 Grade Rams averaged - R2 471
- 31 Seven year old ewes averaged - R3 580
- 121 Full grown ewes averaged - R1 480
- 37 Ewe lambs averaged - R1 188

*Figure 1.10   Meatmaster ram CRC 7203 sold for R34 000 in 2010*

Figure 1.11  Buyers at a Meatmaster sale proving a high level of interest in the new breed

At the National Meatmaster sale on 11 August 2010 held at Hopetown, 65 Meatmaster rams were sold for an average price of R 6 123 and 80 ewes were sold for an average of R 2 568. The top price ram was sold for R 30 000 and the highest price for a ewe was R 12 500.

At the sale on 8 September 2010 following completion of the veld-ram tests at Griekwastad, the ram CRC 9077 was sold for R 40 000 and commercial farmers showed considerable interest in the new breed. New record prices for Meatmaster breeding rams were obtained at the production sale held at La Rochelle on 24 February 2011. The ram CRC 7224 was sold for R82 500 and the ram CRC 7417 was sold for R 42 000.

### 1.2.2  Organized actions to disseminate information to establish the new Meatmaster sheep breed

During the period 2005 to 2011 the members of the Meatmaster Breeders' Society were involved in a large number of activities and actions aimed at the country wide establishment of the new Meatmaster sheep breed.

#### 1.2.2.1  Conferences and symposia

- FW Peters presented the new Meatmaster breed to the international delegates of the 6th Global Conference on the Conservation of Farm Animal Genetic Resources during October 2005 and to the South African Society of Animal Science Symposium held at Upington during July 2006. Research results on the Meatmaster were presented at a Zoology colloquium held at the University of Johannesburg during March 2009.
- A member of the Meatmaster society presented the breed to international representatives

of small stock breeds in Brazil during November 2009
- The President of the Meatmaster society, CR Collett presented the Meatmaster breed of South Africa to the International Clean Skin Sheep Conference in Australia during February 2010.

#### 1.2.2.2  Actions taken and events organized to establish the Meatmaster breed (2005-2011)

These activities included inter alia, National sales, exhibitions, farmer's days, general and special meetings of the Meatmaster society and training courses for present, new and prospective Meatmaster breeders.

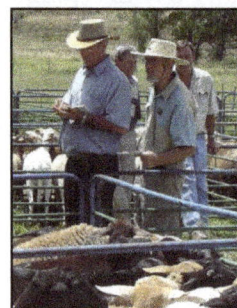

Figure 1.12  Historic first Meatmaster sale at La Rochelle in the Eastern Cape, Clynton Collett and Freddie Peters discussing the catalog (Feb 2005).

**2005:**
- Inaugural meeting of the Meatmaster Sheep Breeder's Society of South Africa. 4 February 2005. Brandfort, Free State.
- First production sale of Meatmaster sheep (500 ewes and 80 rams) and first AGM of the Meatmaster society (12 members) 16 February 2005. La Rochelle, Eastern Cape. (Fig 1.13)
- National slaughter lamb competitions. Meatmasters achieve 9th place amongst 59 group entries and were allotted 22nd, 23rd and 55th places in the single carcass entries amongst 236 entries at Bethulie in the Free State.

*Fig 1.13  The first Council of the Meatmaster Sheep Breeders' Society of South Africa. Clynton Collett (President), Christine du Toit (Vice-President), Freddie Peters (secretary) and Jean du Plessis. (2005 - 2007).*

## 2006:

- Collett farming production sale. La Rochelle. 14 February 2006. Eastern Cape.
- Presentation of Meatmasters to the Prieska farmers association. March 2006. Northern Cape.
- Submission of formal request for proclamation of the new Meatmaster sheep breed and application for the registration of the Meatmaster Sheep Breeders' Society of South Africa to the National Department of Agriculture. Pretoria. 28 April 2006. Gauteng.
- Evaluation of Meatmaster breed standards, performance testing and selection of Meatmaster sheep. La Rochelle. 31 May and 1 June 2006. Eastern Cape. (Fig 1.6)
- Exhibition of Meatmaster sheep at Nampo. Bothaville. 18 to 21 May 2006. Free State.
- Meatmaster society board meeting and 2nd AGM (20 members). Bloemfontein. 8 August 2006. Free State.
- First national sale of Meatmaster sheep (40 rams and 260 stud and commercial ewes from six breeders). Bloemfontein. 9 August 2006. Free State.
- News article. Farmers Weekly. Engineering Africa's ultimate veld sheep. 8 September 2006 Eastern Cape (Burgess, 2006)

## 2007:

- First export of Meatmaster rams to Botswana. Meyerton. 7 January 2007. Gauteng.
- News article. Meyerton ster. Pioneering new sheep breed. 12 January 2007. Gauteng.

- Meatmaster farmers' day presentation. Hopetown. March 2007. Northern Cape.
- Meatmaster society board meeting. La Rochelle. 17 April 2007. Eastern Cape.
- Meatmaster breeders' training courses. La Rochelle. 17 to 19 April 2007. Eastern Cape.
- Meatmaster article in Go Farming. Vol 2. No.1. May 2007. (Alberts, 2007)
- Meatmaster Research Poster prepared and exhibited at the exhibition of Meatmaster sheep at Nampo. Bothaville. 18 to 21 May 2007. Free State. (Appendix C)
- First Meatmaster exhibition and sale at the Pretoria show. 31 August and 1 September 2007.
- Meatmaster society 3rd AGM (28 members) and training courses for beginner members and adjudicators at Meatmaster exhibitions. La Rochelle. 12 and 13 October 2007. Eastern Cape. (Fig 1.14)
- First Cape Meatmaster championships at Loeriesfontein agricultural show. Loeriesfontein. 19 October 2007. Northern Cape.

## 2008:

- Collett farming Meatmaster production sale. La Rochelle. 12 February 2008. Eastern Cape.
- Meatmaster breed promotion and training at Hopetown show. 15 and 16 February 2008. Northern Cape.
- Meatmaster society board meeting. Restructuring of the Meatmaster board and reallocation of responsibilities of board members. Ramsem Bloemfontein. 13 March 2008. Free State.
- Exhibition of Meatmaster sheep at Nampo. Bothaville. 13 to 16 May 2008. Free State.
- Sponsoring of new Meatmaster promotional folders and fliers. 2008
- First national exhibition and sale at Rustenburg agricultural show. 25 to 28 of May 2008. North West. (van Rooyen, 2009)
- Meatmaster society board meeting, 4th AGM (42 members) and three breeders' training courses. La Rochelle. 21 to 24 August 2008. Eastern Cape. (Fig 1.17)
- Cape Meatmaster championships and sale. Loeriesfontein. 23 and 24 October 2008. Northern Cape.

## 2009:

- Collett farming production sale (100 rams and 400 ewes). La Rochelle. 13 February 2009. Eastern Cape.

Fig 1.14    The Meatmaster Council (2008 – 2009) Clynton Collett (President), Christine du Toit (Vice-President), Freddie Peters, Jean du Plessis, Jan Grobler, Danie Visser, Wiid Roe, Marcel van der Merwe with Riaan Liebenberg and Henk Kearney co-opted.

- Meatmaster society board meeting. Hopetown. 20 February 2009. Northern Cape.
- Meatmaster breed representatives exhibition and national sale. Hopetown. 20 and 21 February 2009. Northern Cape.
- Exhibition of Meatmaster sheep at Nampo. Bothaville. 12 to 15 May 2009. Free State.
- National exhibition and national sale. Rustenburg. 2 and 3 of June 2009. North West.
- Meatmaster breeders' courses for new members of the society. La Rochelle. 18 to 20 June 2009. Eastern Cape.
- Visit to Gellap-Ost research station at request of the National Department of Agriculture in Namibia to advise on Meatmaster sheep breeding. Keetmanshoop. 7 July 2009. Namibia.
- Meatmaster society board meeting and 5th AGM (60 members). Hopetown. 12 August 2009. Northern Cape.
- Meatmaster exhibition and national sale (70 rams and 200 ewes). Hopetown. 13 August 2009. Northern Cape.
- Drafting of Constitution and By-laws (in En-

glish) of the Meatmaster Sheep Breeders' Society of South Africa  (Peters, 2009).
- National exhibition and national sale of Meatmasters. Loeriesfontein. 21 to 23 October 2009. Northern Cape.
- Preparation of a number of articles and contributions by members to the first Meatmaster journal.
- Askham farmers' day presentation of Meatmasters to farmers in the Kalahari.

**2010:**
- Collett farming production sale (71 rams and 189 ewes). La Rochelle. 9 February 2010. Eastern Cape.
- Clean Skin Sheep Symposium. Adelaide. 18 and 19 February 2010. Australia.
- Meatmaster breed representatives exhibition and board meeting. Hopetown. 19 and 20 February 2010. Northern Cape.
- Namibia Meatmaster breed information tour. Grünau, Maltahöe and Aroab. 8 to 12 March 2010. Namibia.
- Research article in Small Ruminant Research on Meatmaster genetics. May 2010. (Peters et al., 2010).
- New Era reports the official recognition of the new Gellapper line of Meatmaster sheep by the Department of Agriculture in Namibia following the report by Peters on the analysis of the Gellapper DNA data. 14 July 2010. Namibia.
- Meatmaster breeders' course, AGM and national sale. Hopetown. 9-11 August 2010. Northern Cape. (Fig 1.18)
- First sale of Meatmaster Veld-rams tested at Griekwastad. 8 September 2010 Northern Cape.
- Netherlea Farmersday. October 2010. Northern Cape.
- National exhibition and national sale. Loeriesfontein. 20-22 October 2010. Northern Cape.

**2011:**
- Meatmaster exhibition. Hopetown. 19 February 2011. Northern Cape.
- Collett farming production sale. La Rochelle. 24 February 2011. Eastern Cape.
- Meatmaster veldram sale. Upington. 4 May 2011. Northern Cape.
- Meatmaster exhibition at Nampo. Bothaville. 17-20 May 2011. Free State.
- National sale and exhibition. Meeting of Northern Meatmaster Club. Rustenburg. 31 May - 2

June 2011. North West.
- Meatmaster breeders' courses and national sale. Hopetown 8-11 August 2011. Northern Cape.

*Figure 1.15 Namibian farmers at a Meatmaster breed information day at Grünau*

### 1.2.2.3 Meatmaster training courses

Meatmaster training courses offered included a junior Meatmaster course, senior Meatmaster course, adjudication of Meatmasters and performance testing.

The following aspects are covered during the courses:
- Procedure to join the Meatmaster Sheep Breeders' Society of South Africa, SA Studbook Association and the National Smallstock Improvement Scheme coordinated by the ARC.
- Breeders membership categories, implementation of stud breeding, birth notification of lambs, performance testing, identification system, contemporary groups, environmental factors affecting performance, index calculation, collection of data, BLUP (Best Linear Unbiased Prediction) values and the interpretation of results to improve profitability of the flock.
- Breed standards for Meatmaster sheep, cull faults, selection norms and recommended production systems (Du Toit et al., 2007).

*Fig 1.16   Meatmaster breeders on a training course at La Rochelle*

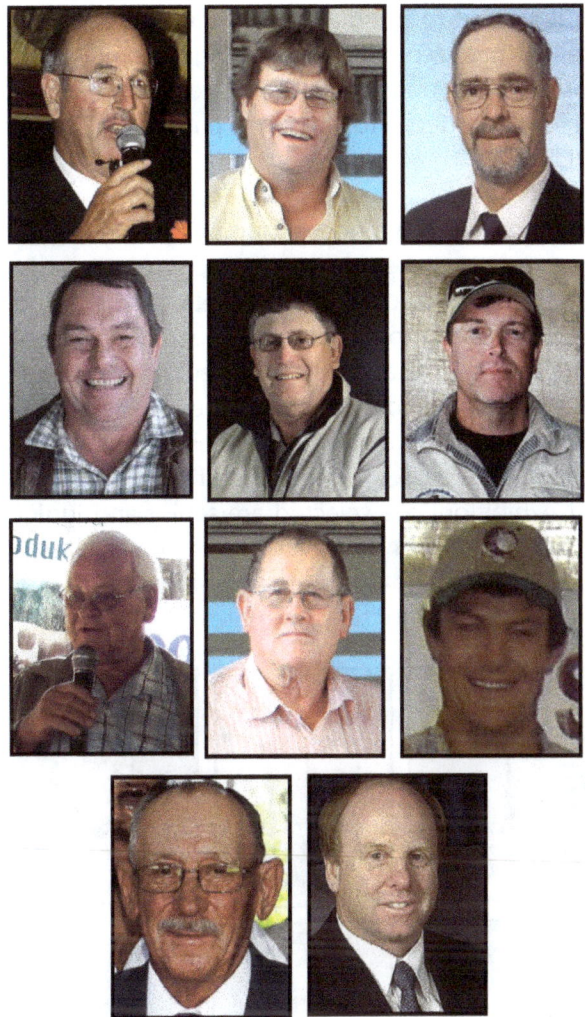

*Fig 1.17   Expanded Meatmaster Council (2009 – 2010): Clynton Collett (President), Jean du Plessis (Vice-President), Freddie Peters, Danie Visser, Jan Grobler, Henk Kearney, Wiid Roe and Dennis Steenkamp with  Riaan Liebenberg and Rassie van der Vyver (co-opted) and Dr Buks Olivier as ARC representative.*

### 1.2.2.4 Meatmaster shows and exhibitions

Bonsma (1983) commented on livestock shows: *"Livestock shows have come to stay, but they should be modified to include a better educational program where the placing of animals is explained to the interested breeders in terms of functional efficiency."*

In order to establish the Meatmaster sheep breed and to assist prospective breeders to become better acquainted with the breed, the Meatmaster society developed a completely new and unique concept towards exhibitions and shows. No competition between sheep or breeders is allowed. Breeders exhibit their sheep and discuss individual sheep with each other and the adjudicators. Each exhibitor is allowed to explain why he decided to bring a specific sheep to show to fellow breeders. Performance figures and other outstanding traits of the sheep are pointed out to fellow breeders and interested parties (Du Toit et al,. 2007).

During the exhibition a number of breed representatives are selected on a consensus basis. There is no fixed numerical quotas applied and two, three or more sheep are selected in each of the four classes as breed representatives, indicating to interested parties what the breeders think they should strive to attain by their breeding programs. The classes in which breed representatives at exhibitions are selected are senior and junior ewes and senior and junior rams. Junior classes indicating sheep up to the age of two teeth and senior classes four teeth or more.

*Fig 1.18    The New Council (2010 – 2011): Danie Visser (President), Henk Kearney (Vice-President), Clynton Collett, Jean du Plessis, Dennis Steenkamp and Wiid Roe with Riaan Liebenberg,  Freddie Peters, Rassie van der Vyver and Johnny Morrisson co-opted members and Dr Buks Olivier as ARC representative.*

## 1.2.3 Proclamation of the Meatmaster sheep as a recognised new sheep breed

An absolute crucial part in the achievement of the establishment of the Meatmaster sheep breed was to obtain official recognition of the breed. Prior to the official proclamation of the breed as a developing breed, the Registrar of Animal Improvement stated that the Meatmaster Breeders' Society was 'acting against the law by promoting a breed that was not officially proclaimed in terms of the Act. An order was issued by the Registrar forbidding any form of advertising of the Meatmaster sheep and the Meatmaster Breeders' Society was expected to cease its activities. Breeders were then demoralised and despondent and confidence in the future of the Meatmaster sheep declined rapidly. Only the perseverance of a small number of breeders prevented total disillusionment.

In terms of section 2 (2) of the Animal Improvement Act (1998) it is stipulated that in the case of a new kind of animal or a new breed of such kind of animal to be imported into or to be bred in the Republic, the Minister shall make such declaration after considering the request, taking the international law into consideration and after considering comments received in response to an invitation by the registrar to interested persons to comment on a proposed declaration that had been published in the Gazette at least 30 days prior to such declaration.

The Meatmaster breed is a new landrace breed bred in the Republic of South Africa. In order to establish the Meatmaster as a new breed it was of paramount importance to have the breed formally proclaimed in terms of the Act. The Meatmaster breeders' society could only be registered in terms of section 8 (7)(a)(i) of the Act after the Meatmaster breed was officially recognised. There were no clearly defined guidelines and criteria specified in terms of which it was possible to make a realistic assessment as to when such proclamation would be made. Following numerous personal discussions with senior officials of the Department of Agriculture it was concluded that it would be necessary to lodge a formal application to the Registrar of animal Improvement in which sufficient information regarding the breed standards, numbers of animals bred by at least seven breeders, evaluation of agro-economic aspects of the breed and a constitution of the prospective breeders' society which conformed to the stipulations of the Act, should be submitted.

### 1.2.3.1 Official recognition of the Meatmaster sheep breed

Following the application, motivation and numerous discussions with the registrar of animal improvement (Appendix B1), the Meatmaster sheep breed was officially recognised as a newly developing breed by the publication of a Government Notice in Government Gazette No. 29898, dated 25 May 2007 (Appendix B2). The Meatmaster sheep breed thus became one of the nineteen sheep breeds in the Republic of South Africa which are officially recognised by the Department of Agriculture. This was undoubtedly a major achievement. The official recognition of the Meatmaster sheep breed immediately drew much attention from the general farming community in the Republic of South Africa and numerous new members joined the Meatmaster breeder's society taking the popularity of the breed to a new level.

### 1.2.3.2 Official recognition of the Meatmaster Sheep Breeders' Society of South Africa

The registrar of animal improvement informed Meatmaster breeders on 20 August 2008 that the Meatmaster Sheep Breeder's Society was approved by the Minister of Agriculture and that a notice to that effect would be published in the Government Gazette as soon as SA Studbook supplied the Department of Agriculture with the required serial numbers. Soon afterwards the President of the Meatmaster breeders' society was informed by SA Studbook that the Meatmaster breeders' society was now formally recognized. This marked the beginning of a totally new era for the Meatmaster breed and Meatmaster sheep breeders in South Africa. The small number of Meatmaster breeders was up to this point in time constantly being criticized for merely trying to market a number of crossbred sheep as a new breed.

During the following period of about 18 months Meatmaster breeders were inundated by calls for information on the new breed and requests for breeding material. The existing Meatmaster breeders could not nearly satisfy the sudden demand. More exhibitions, national sales, information days and training courses had to be offered and the number of registered members of the Meatmaster Sheep Breeders' Society of South Africa increased from 20 to more than 60 members (Fig 1.19), thereby becoming the 4th largest breeders' society amongst the 10 meat and dual purpose sheep breeders' societies in South Africa. The rapid growth in the number of Meatmaster sheep and Meatmaster sheep breeders in the country led the National Department of Agriculture to upgrade the Meatmaster sheep breed and to reproclaim the Meatmaster as a regularly introduced breed. (Appendix B3)

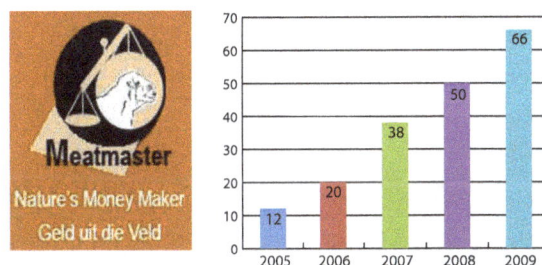

Fig 1.19   Growth of the Meatmaster breeders' society (2005-2009)

### 1.2.3.3 Official upgrading of the Meatmaster sheep breed

Government Gazette no. 32601 dated 2 October 2009, reproclaimed the Meatmaster sheep as a *"locally adapted and regularly introduced breed"* alongside well known sheep breeds such as the Suffolk, Ile de France and Merino land sheep (Appendix B3).

Following the reproclamation of the Meatmaster breed, international interest in the new breed attained new heights and several applications from prospective Meatmaster breeders abroad were received by the Society to accommodate them as registered Meatmaster breeders.

The Meatmaster sheep breed is expected to continue to grow in numbers and popularity as the economic value of the breed becomes more apparent. The Meatmaster sheep breed is in fact a locally developed landrace breed and there is no reason why the breed should not shortly be proclaimed as such alongside landrace sheep breeds like the Dorper, Afrino and SA Mutton Merino.

*Chapter 1*                                                                    *Introduction*

### 1.2.3.4   *Official recognition of the Gellapper line of Meatmaster sheep in Namibia*

New Era reported on 14 July 2010 that the new Gellapper line of Meatmaster sheep was now officially recognized by the Namibian Department of Agriculture following the report by FW Peters on the analysis of the Gellapper DNA data (Chapter 4).

**14**

# Jerued Meatmasters

## Nature's Money Maker

Koppies, Free State

Examples of ewe's with their 4th lamb after lambing the first time at 13 months.

Meatmasters is the easiest breed to farm with, low on maintenance, highly fertile, exceptional grazers, good mothers that rear their lambs adequately and masculine, fertile rams with excellent serving ability.

Die aanpasbaarheid van hierdie diere is ongelooflik. Met net die veldweiding hier by ons in die Vrystaat, presteer hulle voortreflik.

**For any enquiries contact:**

Koetoe Botha at cell: 083 626 6698 or email: koetoeb@actionford.co.za
Folos Botha at cell: 072 112 8732 or email: folos@jeruedmasters.co.za

# Die Bult
## Meatmasters

Danie Visser • Edenville, Kenhardt • Cell: 083 390 0860 • Email: dhvisser@telkomsa.net

# Chapter 2  Phenotypic description of the Meatmaster sheep, motivation of breed standards and production conditions

## 2.1  The Meatmaster - A phenotypic description

Breed reference name:      MEATMASTER
Local synonyms:            Bosrander, Damara Crossbred Sheep.
International synonyms:     Damper, Droughtmaster, Gellapper.
Breed:                     Composite Landrace

### 2.1.1  Photographs

The photographs in figures 2.1 - 2.3 are good examples of the average Meatmaster ram, ewe and veld conditions.

Figure 2.1   Meatmaster ram

Figure 2.2   Meatmaster ewe with lambs

Figure 2.3 Typical Meatmaster sheep in natural veld conditions for which this breed is well adapted

### 2.1.2  Breed distribution

The breed is already established in the Northern Cape, Gauteng, Southern Free State, North Eastern Cape, Central Northern Provinces and the Drakensberg area of Kwazulu-Natal. Meatmaster breeding rams were recently exported to Namibia and Botswana. A population of Meatmasters is also being bred in Australia from the initial crossing of Dorper and Damara sheep recently imported from the Republic of South Africa.

### 2.1.3  Breed physical characteristics

#### 2.1.3.1  Coat

Colour: Multi-coloured, unicoloured or pied white, black, brown, tan, black and white patched, black and brown, dark-brown, large variety of colour patterns.

Description:  Hair and fluffy Wool
Hair: Hair type: Straight
Hair length: Short
Wool: Fluffy wool in winter that sheds itself in summer

#### 2.1.3.2  Head

Face profile: Convex
Muzzle colour: Pigmented
Throat ruff: Small or Absent
Toggles: Common
Horns: Present tending to be polled or polled
Horn number: male 2 if present
Horn number: female 2 if present
Horn shape: Curved
Horn length: Short in males, short in females
Horn orientation: Lateral
Ears: Ear size: Medium
Ear orientation: Drooping

#### 2.1.3.3  Body

Body frame: Medium
Back profile: Slightly Hollow
Rump profile: Sloping
Legs: Long
Hooves: Dark

*Tail:* Tail length: Long
*Tail thickness:* Thin

### 2.1.4  Natural production environment

*Climatic condition:* Found in areas with a median annual rainfall ranging between 126 and 600 mm, a median maximum daily temperature of 36°C in the hottest months and a median minimum daily temperature of -3°C in the coldest months.

Figure 2.4   Two typical production environments, Meatmasters in the Eastern Cape (top) and Meatmasters on the Highveld (bottom).

### 2.1.4.1  Vegetation type

Arid Karoo, Kalahari thornveld with mixed bushveld and thornveld in Gauteng and the Northern Province.

### 2.1.4.2  Production system

Mainly extensive commercial systems. Increasingly popular in bushveld areas.

*Adaptations:* Adapted to hot arid environments. The animals walk long distances in search of food and water and can produce and reproduce on minimal grazing. They are said to defend their lambs from attacks by small predators such as jackals.

### 2.1.4.3  Health management

Minimal input in arid areas.

### 2.1.5  Distinguishing characteristics

Male with prominent occipital chest, wattles are common.

## 2.2  Meatmaster breed standards

### 2.2.1  Aim of this part of the chapter

The aim of this part of the chapter is to describe the Meatmaster breed standards based on agro-economic principles. Well defined breed standards is a pre-requisite for breed recognition and defines the criteria to be met by individual Meatmaster breeders. Not all of the Meatmaster populations have the same breed compositions but selection is based on the same breed standards.

Bonsma (1983) commented as follows on breed standards: *"It is my considered opinion that some breed societies have fixed ideas about what the ideal type of livestock is for their particular breed. They have made their breed standards so concrete that no modification can take place and they have reached a point where they cannot select desirable variants; hence they cannot improve their breed. These fixed idols of clay or brass are nothing but the prejudiced visions of men who really worship an image that has not been measured by performance testing nor functional efficiency. Prejudice and conservatism are the main obstacles which in many instances prevent the improvement and progress of certain breeds."* Estimations of genetic parameters in the Glen Dorper flock indicated that production traits were moderately heritable and that genetic progress was feasible. No mentionable genetic progress was however found for any trait during the period from 1982 to 1990, when selection was primarily based on breed standards (Schoeman et al., 2010).

Meatmaster breed standards define a completely new and distinct breed of sheep which notably ignores petty conformation issues and is dedicated and committed to the achievement of the important agro-economic factors that aim to keep costs as low as possible and maximise income. Breed standards should be reviewed from time to time in order to facilitate genetic progress in line with research results obtained by experienced animal breeding scientists. Meatmaster sheep are bred exclusively in veld conditions in order to identify the most adaptable and productive breeding stock capable of producing in adverse ecological conditions. A preliminary attempt will be made to quantify the value of the breed standards in terms of estimated savings or added value. These estimations are not yet scientifically supported but are included to serve as a guideline to future researchers who may wish to evaluate the true monetary value resulting from these breed standards. The Meatmaster breed standards are based on the philosophy that a breed standard should either add value or decrease operating costs in order to improve profitability.

### 2.2.2 Colour and skin pigmentation

*Figure 2.5    Colour variations in Meatmaster sheep are unlimited*

Any colour or combination of colours is acceptable, but good skin pigmentation around the eyes and over the ears is essential.

*Agro-economic motivation:* Good skin pigmentation provides natural protection against sunburn in harsh climatic conditions. The decrease of skin problems associated with weak pigmentation will consume less time, treatment and effort.

*Quantification estimate:* Dorper skins are well known to classify as Cape Glover quality skins (Du Toit, 2007), which realise a premium in the hide and skin market. Du Toit (2007) quoting Anderson (1856) referring to the coat colour of the Damara

sheep states that the greatest peculiarity of these animals is their colour, which is of every hue and tint. The Meatmaster sheep which has both of the former breeds as parent breeds has a Cape glover quality skin presenting itself in innumerable colour patterns in variations of black, white, brown, grey and tan.

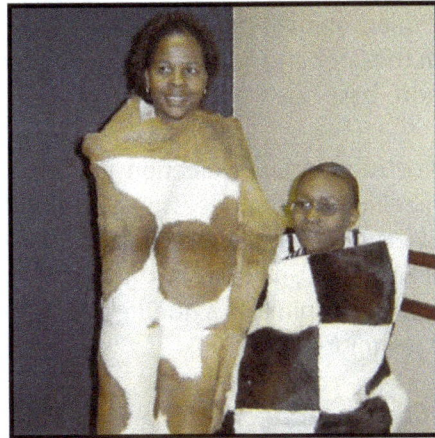

*Figure 2.6    Examples of hair sheep skin products*

Du Toit (2007) furthermore presents attractive examples of skin products, a jacket and briefcase from Damara leather, a karos or blanket stitched from four skins, different ottomans and a carpet. Similar products can be made from Meatmaster skins thus adding additional value. At an estimate of 1% of skins that can realise an additional R300 in curios value this will add R3.00 per unit in value and by saving 1% of the sheep flock from being treated for ill pigmentation problems at R300 per unit treated a saving of R3.00 per unit will result. This breed standard will then add an estimated R6.00 per unit in value resulting from additional income and cost savings.

### 2.2.3 Head and horns

*Figure 2.7 (a)    Ewe*          *Figure 2.7 (b)    Ram*

Ewes must be feminine and rams masculine. Horns are acceptable in both sexes, but a preference for polled animals is actively encouraged.

*Agro-economic motivation:* The general appearances of the sexes should complement its respective functionalities. Larger physical built in the rams will enhance meat production income from ram lambs and wethers and smaller built in the ewes will increase the carrying capacity of the number of breeding ewes due to lower maintenance requirements (Snyman and Herselman, 2005). Polled animals cause less injury to other sheep in the flock and reduce the risk of infections at the horn base due to injuries.

*Quantification estimate:* Selection for polled animals will eliminate the necessity of treatment of horn-based infections due to injury. A nominal value of R1.00 per unit cost saving is estimated on the basis of one animal in a hundred to be treated at R100 per injury.

The coat should consist mainly of short, glossy hair of any colour or colour pattern, with an undercoat of fine, fluffy wool. Meatmasters should shed the coat covering in summer but retain enough fluffy wool for protection during winter. Shedding their coats naturally is essential.

*Agro-economic motivation:* No shearing costs would thus be incurred, no tick or blowfly treatment is necessary and there is no after shearing weather risk.

*Quantification estimate:* Natural shedding of the coat covering will save shearing costs and prevent seed penetration through the skin which causes slaughter losses. No treatment for blowfly strike will contribute to operating cost savings. Estimated at R8.00 per unit (GADI, 2010).

### 2.2.4 Coat covering

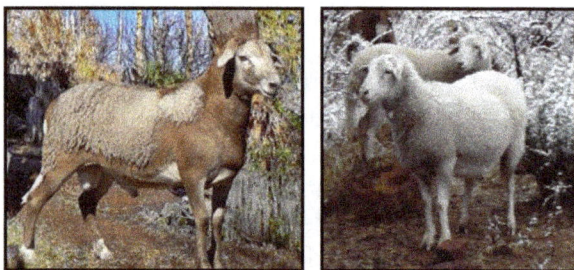

*Figure 2.8   Coat of fine fluffy wool in winter*

| Wool related costs determined for a 1000 ewe flock in the UK. (Walker, 2010) | | |
|---|---|---|
| **Operation** | **Notes** | **Pence per ewe (UK £)** |
| Shearers | Per ewe(incl. hoggs & rams) | 123 |
| Labour at Shearing | 2 casuals for 2 days (7.5/hr) | 24 |
|  | Shepherd for 4 days (10/hr) | 32 |
| Lamb growth loss at Shearing | 150g/day for 3 days | 45 |
| Crutching ewes | Per ewe | 45 |
| Labour at Crutching | Shepherd for 2 days | 16 |
| Dagging 25% of lambs | At L% of 150 @ 45p | 17 |
| Labour for Tail docking | Total time 6 hours | 5 |
| Lamb growth loss at Tail docking | 150g/day for 3 days | 45 |
| Blowfly control | Pour-on twice @ 50p | 100 |
| Labour at blowfly control | Shepherd for 2 days | 16 |
| Lamb growth loss at blowfly control | 150g/day for 1 day | 15 |
| No deaths of cast ewes | 5 per 1000 ewes | 25 |
| Reduction in shepherding for cast ewes | 30 days at 1 hour per 1000 ewes | 30 |
| **Total Cost Saving** |  | **538** |
| Wool income foregone | 2.72kg @ 21p/kg | 57 |
| **NET SAVING FROM HAIR SHEEP** |  | **481** |

## 2.2.5  Conformation and legs

Figure 2.9   Functional body conformation

Meatmasters must be of average size with a functional, efficient body conformation and well-placed legs, so that they can walk long distances.

*Agro-economic motivation:* A functional conformation will ensure that the sheep can walk long distances easily, will graze efficiently and will not lie down around the water troughs. Actively moving sheep will also be less prone to infestation by external parasites.

*Quantification estimate:* The conformation which enhances easy lambing accounts for the low lamb mortality rate observed in Meatmaster sheep. Lambing difficulties also increase the mortality rate amongst ewes. At an estimated saving of one in one hundred and fifty ewes valued at R900 per ewe the saving amounts to R6.00 per unit.

### Body conformation of rams

Figure 2.10   Ram conformation

Rams should be masculine in general appearance. They may have horns but should preferably tend to be polled. Rams should generally be of larger build than ewes.

*Agro-economic motivation:* Polled rams will cause fewer injuries to other sheep in the flock and will be less prone to infection at the base of the horns which may be caused by injuries. Larger rams will normally produce heavier slaughter lambs and wethers.

### Body conformation of ewes

Figure 2.11   Ewe conformation

Ewes should have a general feminine appearance. They should be of average size and possess the herd instinct to stay with the flock.

*Agro-economic motivation:* Ewes of average size have been proven to be the most functionally efficient breeding stock (Schoeman, 1996). Ewes should adhere to the flock in order to minimise the loss of lambs to predators.

## 2.2.6  Hindquarters

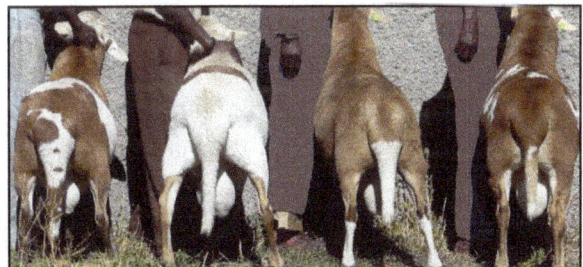

Figure 2.12   Well fleshed hindquarters

Hindquarters should be well fleshed with no fat localisation on the rump.

*Agro-economic motivation:* The carcass quality should be acceptable to the market and should meet market standards (Hofmeyer, 2001) in order to realise more valuable meat production income. One of the characteristics of breeding goal traits must be desirable economic value, either as a marketable commodity or as a means of reducing production costs (Tibbo, 2006). The Meatmaster is not a fat tailed type of sheep against which the market tends to discriminate.

*Quantification estimate:* Well fleshed hindquarters is a general breed standard for mutton sheep breeds, therefore no additional value can be assigned to this breed standard relative to most other breeds. When compared to the Damara parent breed, additional value is however added in terms of market acceptability as well as a better carcass grading.

### 2.2.7 Testicles

Figure 2.13   Well formed testicles

Testicles should be well formed with minimal abnormalities.

*Agro-economic motivation:* Functional efficiency in this regard will ensure optimum productivity. The Meatmaster breed has minimal abnormalities.

*Quantification estimate:* Although the Meatmaster breed has minimal abnormalities in this respect, no additional monetary value is added to this breed standard relative to other breeds.

### 2.2.8 Tail

Figure 2.14   Ideal Meatmaster tail

The tail should be well attached and preferably not longer than the hock. It should be a neat wedge-shape or uniformly thin, with only a moderate covering of fat. Docking of the tail should never be necessary.

*Agro-economic motivation:* Ensures reduced labour costs and minimum lamb losses due to stress and post docking infections. Walker (2010) estimated lamb growth loss at tail docking at 150g/day for 3 days.

*Quantification estimate:* The tail itself provides additional carcass weight, estimated at 1kg at R25 per kg and at a cost saving of tail docking material and equipment as well as labour costs associated with tail docking. A saving of R1.00 per unit is estimated in this regard. A total additional value of R26 per unit is estimated for this breed standard.

### 2.2.9 Herd instinct

Figure 2.15   Meatmasters keep closely together due to their strong herd instinct

A strong herd instinct is vital, culminating in a visibly strong togetherness of the flock - it simplifies management in dense bush or mountainous terrain and prevents small groups wandering away on their own or getting through fences. Sheep with a well-developed herd instinct protect themselves and their lambs from predators (personal observation).

*Agro-economic motivation:* Reduces labour costs and losses considerably because larger flocks can be looked after with ease. The strong sense of direction and location of the Meatmaster ensures that they return to base regularly and naturally.

*Quantification estimate:* The herd instinct of a close togetherness limits the loss of animals due to predators. At an estimated life saving of two lambs per hundred, this breed standard accounts for R12 per unit of additional value.

## 2.2.10  Conclusions

The Meatmaster breed requirements confirm that there is no sheep breed which has even remotely the same breed standards.

The Meatmaster doesn't have to conform to any norms prescribing body shape - the breed standards don't specify what the ears should look like, how the neck and shoulder should be attached, or the shape of the forequarter or any part of the body. It is simply states that the Meatmaster sheep should have a good, functionally efficient body conformation.

The total additional value estimated as a direct result of the breed standards amounts to R56 per ewe in the breeding flock.

Further research to evaluate the true monetary value of the Meatmaster breed standards is recommended.

## 2.3  Production conditions and quality assurance of Meatmaster sheep

The generally accepted production conditions for Meatmaster sheep is that they should be able to produce profitably in veld conditions without supplementary or additional feeding. Meatmaster breeders will use different management systems depending on the local conditions dic-

tated by diverse agro-ecosystems. As an example the management system followed by CR Collett (Collett, 2007) in the North Eastern Cape is cited as a guideline.

### 2.3.1  Example of a Meatmaster management system in the Eastern Cape

All Meatmasters run on a multi camp system and change camps every two weeks. All ewes are mated, lamb and rear their lambs, including all twins, without exception in the veld under natural conditions. All lambs are also weaned just onto natural veld. Collett encourages twins with his selections but ewes must be able to rear their lambs in the big groups under these natural conditions.

*Figure 2.16   Meatmasters on a multi-camp system*

All stud sheep are recorded but for selection of the flock sheep the under mentioned programme is applied:

- All ewe lambs not lambing at 13 months are discarded.
- All ewes not in lamb at lambing time are culled.
- Before lambs are weaned a wet dry test is done on all ewes by way of removing all lambs from their mothers for the night and observing the udders of the ewes the following day, ewes not in milk will not have swollen udders and are culled from the flock because these ewes must have lost their lambs.
- When 85% of the lambs have been sold and only the weakest 15% are left, a wet dry test is again performed to find the mothers of this 15%. These being the weakest mothers are also taken out for sale.
- All rams also spend their entire life just on the open veld. There is no feeding before mating or for the sale of rams. It has to be stressed again that no feeding of any kind ever occurs on any of their Meatmaster breeding sheep.

Only old ewes not fit for further breeding will be fattened for slaughter if necessary. The Meatmasters will only have rock salt available to them throughout the year as a supplement. This management system ensures that only the best breeding material, capable of performing under these adverse conditions is kept for breeding purposes.

All Meatmaster Stud Sheep are recorded on the SHEPHERD Sheep Program. To select the most profitable breeding ewes, data recorded is analysed to provide the following important information:

• Age at first lambing
• Inter lambing period
• Death of lambs from birth to weaning
• Weight of lambs at 100 days
• Weight of lambs at 270 days

Best linear unbiased prediction (BLUP) calculations are subsequently performed by the ASREML program (Gilmour et al., 2000) and imported into the SHEPHERD program. The index values are then used to select the most profitable ewes and rams. Lambs are linearly scored for covering, condition, head, pigment, width and muscling, body length, depth of body, shoulder height, conformation and body size to ensure that all Meatmaster sheep meet the breed standards. BLUP calculations are also used to select young Meatmaster sheep for production potential (Olivier, 2007).

Performance testing is applied to ensure that only the most profitable sheep are selected for breeding purposes.

### 2.3.2  Quality assurance through veld-ram testing. First Meatmasters on veld-ram tests.

Meatmaster ram lambs were entered for veld-ram tests for the first time in the vicinity of Vrede in the Free State during November 2009. Meatmaster breeders from Mpumalanga and Gauteng enrolled ram lambs for the test. Ram lambs from the Dohne Merino, a wool sheep breed and ram lambs from the SA Mutton Merino, a dual purpose breed

were also enrolled. Interim mean average daily gain (ADG) values (± s.d.) recorded for the breeds were ADG1: Dohne Merino (n=20) ADG1=94±54 g/day; SAMM (n=59) ADG1=106±51 g/day, Meatmaster (n=17) ADG1=132±54 g/day and for Dohne Merino ADG2=66±44 g/day; SAMM ADG2=87±38 g/day and for Meatmaster ADG2=106±45 g/day. There were no statistically significant differences for ADG1 values and for ADG2 there was only a significant difference between the Meatmaster and the Dohne Merino breeds. Famacha© chart scores were also recorded and interim mean Famacha scores (± s.e.) were for Dohne Merino (n=20) F© = 4.05±0.17; SAMM (n=59) F© = 4.40±0.07 and Meatmaster (n=17) F© = 3.9±0.2. Meatmaster scores were significantly lower than the value for SAMM (P<0.05). A validation study of the Famacha© system for clinical evaluation of anaemia due to Haemonchus contortus was conducted on sheep farms in the summer rainfall region of South Africa (Reinecke et al., 2009).

Meatmaster rams were also enrolled for the first time for veld-ram tests at Griekwastad in the Northern Cape during 2010. The ARC recorded data at the Griekwastad tests which also included Dorper rams. Detailed results were not yet released.

Fig 2.17    Veld-ram tested Meatmaster rams at Griekwastad

The Meatmaster breeders' society will continue to encourage veld-ram tests for the breed in order to ensure viable and quality tested breeding stock for commercial farmers.

# Olive Branch
## Meatmasters

*Senior Champion Ram at Rustenburg show 2011*

**Contact Rachel Fouché**
*Cell: 084 240 5269*
*Tel:018 571 3804*
*E-mail: rachelfouche@hotmail.co.za*
*Web: www.foucheboerdery.co.za*

**Fouché Boerdery**
*Lichtenburg, North West*

*Now registered on Studbook.*

# MEINFRED

## MEATMASTERS

Freddie Peters • +27 16 365 5350 (tel/fax) • +27 83 521 3990 • freddiepeters@gmail.com

Ons teel Meatmasters sedert 1994. Aanvanklik geteel uit Damara en Ile De France kruisings. Later is ramme gebruik uit die Dorper/Damara basis kuddes. Hoofsaaklik uit Clynton Collett se stoet. CRC 0025 (Top prys ram in 2005), CRC 4034 en CRC 6165 (Seun van CRC 4403, top prys ram in 2008).

Ons het Meatmaster Ramme en van tyd tot tyd groepies van 1 Ram en 5 Ooie beskikbaar.

Meinfred Meatmasters loop op die Hoëveld langs die Suikerbosrand Natuurreservaat naby Meyerton in Gauteng en is gewoond aan jakkalse, rooikatte en bebosde berge.

Algeheel op veld geteel met slegs klipsout lek. Lam en speen op die veld.

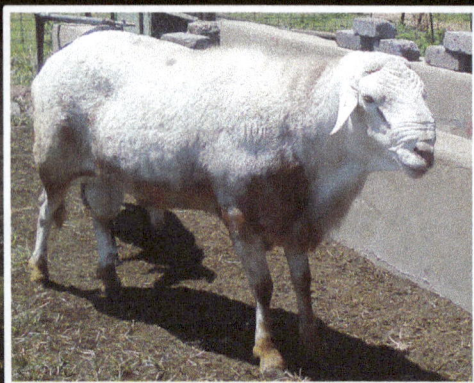

# Chapter 3  Meatmaster breed creation programs

## 3.1  Introduction

A number of different breed creation programs were implemented during the course of the initial development of the Meatmaster sheep. Information on breed creation programs implemented by different breeders was collated for use by future breeders or researchers. Although the breed creation programs were not scientifically designed or based on theoretical animal breeding principles the objective of the author with this chapter is to document all the information on the breed creation programs followed to establish composite Meatmaster sheep populations. The Meatmaster breed is a composite breed with different breed combinations in some populations of the breed.

*"A composite breed is a breed made up of two or more component breeds and designed to benefit from hybrid vigor without crossing with other breeds."* (Bourdon, 2000). Pure composite systems can produce considerable hybrid vigor. When two-breed F1s are mated to produce F2s, half of F1 hybrid vigor is lost but half remains in the F2, F3 and subsequent generations. The amount of hybrid vigor retained depends on the number and proportions of component breeds in the composite. According to Bourdon (2000) intelligent crossbreeding generates hybrid vigor and breed complementarity, phenomena that are important to production efficiency. Bourdon (2000) assumes hybrid vigor to be linearly related to heterozygosity and retained hybrid vigor is based on the dominance model for hybrid vigor. Hybrid vigor in a composite breed will be retained over time if inbreeding is kept to a minimum.

Bourdon (2000) provides a formula for the prediction of retained hybrid vigor:

$$\%R\hat{H}V = \left(1 - \sum_{i=1}^{n} p_i^2\right) \times 100$$

Where $p_i$ is the proportion of the ith breed in a composite made up of n component breeds. Applied to a two-breed composite (50%A and 50%B) retained hybrid vigor amounts to 50%.

Meatmaster populations vary in the %R$\hat{H}$V in accordance with the variety of breed creation pro-grams applied. The differences in breed compositions of Meatmaster populations can be utilized to retain high levels of hybrid vigor in the composite Meatmaster breed.

In order to create initial genetic diversity, the Meinfred Meatmaster stud was initiated by the crossing of five unrelated bloodlines of veld adapted Ile de France stud ewes and three unrelated Damara stud rams. The Damara rams originated from the Kaokoveld in Namibia, the North Eastern Cape and the Northern Province. The Ile de France stud was established on the Schoongezicht farm near Kareekloof, Meyerton during 1982, at least 10 years before cross breeding was commenced.

Other Meatmaster populations were established by different breed creation programs but the Damara was included in all of them as a parent breed. Breed creation programs by which Meatmaster populations were initiated included the Dorper, SA Mutton Merino, Van Rooy and Wiltshire Horn as parent breeds. In Namibia a Meatmaster population was also bred from Dorper and Damara sheep but with different percentages applied to the composition. It was subsequently established that in Australia breed creation programs were initiated where the Suffolk and Damara were included as initial parent breeds. Subsequently, the Dorper, Van Rooy, Wiltshire Horn and Speckled Persian breeds were also introduced into their breed creation programs (Russel, 2010).

## 3.2  The parent breeds

Meatmaster parent breeds which were included in breed creation programs are the Damara, Dorper, Ile de France, SA Mutton Merino, Van Rooy, Wiltshire Horn and Suffolk sheep breeds. A brief description of these breeds is given in order to promote a better understanding of the Meatmaster characteristics which originated from the parent breeds. Schoenian (2009) and Campbell (1995) provide concise descriptions on a large number of sheep breeds.

### 3.2.1 The Damara

Figure 3.1   Damara sheep

The Damara originated from the Hamites of Eastern Asia and Egypt and moved down to the present day Namibia and Angola. For many years, the sheep were in an isolated region of Namibia and thus remained free of influence from other breeds. Damara sheep can survive in a harsh environment and under poor nutritional conditions.

Research has indicated that up to 64% of the diet of the Damara sheep can consist of browsing material (du Toit, 2007). This places the Damara in the same feeding category as goats. Hair is mostly short with a tendency to a fine layer of woolliness developing under the hair during winter. The sheep has a wide range of colours which are all equally acceptable and desirable. With the exception of the tail and the rear rump, there is no visible localized fat.

*Breed categories:* hair (meat), fat-tailed
*Distribution:* Africa, Australia, New Zealand, Canada

### 3.2.2 The Dorper

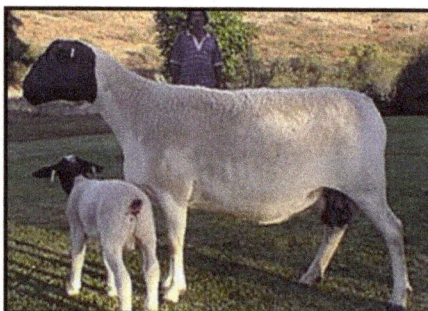

Figure 3.2   Dorper ewe and lamb

The Dorper breed is numerically the second largest breed in South Africa. Their popularity has spread to many countries throughout the world, including the United States in 1995. The Dorper as

developed in South Africa in the 1930's, through crossing of the Blackheaded Persian ewe (a native fat-tailed sheep) with the Dorset Horn (a mutton breed).

The breed has a characteristic black head (Dorper) or can be all white (White Dorper). Dorpers are a well-proportioned breed with heavy muscled hindquarters. Their skin covering is a mixture of hair and wool. The Dorper has a thick skin which is highly prized and protects the sheep under harsh climatic conditions. It is the most sought after sheepskin in the world.

*Breed categories:* hair (meat)
*Distribution:* Worldwide

### 3.2.3 The Ile de France

Figure 3.3   Ile de France sheep

The Ile de France is the result of crossing the English Leicester and the Rambouillet. The breed was originally known as the Dishley Merino. The breed is widespread in France and was introduced to Great Britain in the 1970's. The breed is wide and thick set. Both sexes are polled. The Ile de France is widely used throughout the world as a terminal sire for quality lamb production.
*Breed categories:* meat, medium wool
*Distribution:* Worldwide

### 3.2.4 The SA Mutton Merino

Figure 3.4   SA Mutton Merino – a dual purpose breed

Originally known as the German Mutton Merino, the first ten ewes and a ram were imported to South Africa from Germany in 1932 by the Department of Agriculture for a breed creation programme. Through selection for a better wool quality and conformation, the uniqueness of the South African breed was recognised in 1971 when the breed name was changed to the SA Mutton Merino. The SA Mutton Merino is a dual purpose mutton-wool sheep (80:20 mutton to wool), originally bred for its high adaptability to all farming regions in South Africa. The breed was developed to produce a heavy slaughter lamb at an early age as well as good quality wool.

The breed has been used, as a sire line, in the development of four South African landrace breeds, namely the Dohne Merino, the Afrino, Dormer and the Vandor.

A strong well-muscled polled sheep with an excellent conformation and balance. It is a large framed breed with a fleece of pure white wool - free of kemp and coloured fibers.

*Breed categories:* meat, medium wool
*Distribution:* South Africa, Australia

### 3.2.5   The Van Rooy

*Figure 3.5    Van Rooy*

In 1906, Senator JC van Rooy, of the farm Koppieskraal in the Venterstad district of the Free State, started experiments to propagate a breed of sheep for slaughter lamb production: The requirements he set for this breed, were threefold: 1) the breed had to be strong and hardy to cope with regular droughts; 2) it had to be fertile in order to maintain a high percentage of production; it had to have an excellent conformation.

With these aims in mind he made use of a white *"Blinkhaar Ronderib Afrikaner"* ram and eighty

Rambouillet ewes. With the progeny of these the principle of inbreeding, coupled with severe selection, was applied. Later on, a polled Wensleydale ram was introduced in an effort to improve conformation. The present day Van Rooy sheep is still run mostly in the arid areas where survival and reproduction on natural grazing is essential for the economic production of meat.

*Breed categories:* hair (meat), fat-tailed
*Distribution:* South Africa, Australia

### 3.2.6   The Wiltshire Horn

*Figure 3.6    Wiltshire Horn*

The Wiltshire Horn is an ancient British breed from the Chalk Downs region of England. Reaching large numbers during the 17th and 18th centuries, they became almost extinct by the beginning of the 20th. They are currently attracting attention for their lack of wool and no need for shearing, as well as their ability to pass on their vitality and quality meat in a cross-breed creation program. Both rams and ewes are horned. The Wiltshire Horn is classified as a *"rare"* breed by the American Livestock Breeds Conservancy.

*Breed category:* hair (meat)
*Distribution:* Worldwide

### 3.2.7   The Suffolk

*Figure 3.7    Suffolk*

Suffolks are found throughout the world's sheep producing countries. In the United States, they are by far the most popular pure breed of sheep, accounting for more than fifty percent of purebred sheep registrations. In the British Isles, they are the leading terminal sire breed. The Suffolk breed originated almost 200 years ago on the rugged south-eastern coast of England, the result of crossing Southdown rams and Norfolk Horn ewes. Originally, they were called Southdown Norfolks or just *"Black faces."*

They are a large breed with a distinctive all-black head and legs that are free of wool. Suffolk lambs grow fast and yield heavy carcasses.

*Breed categories:* medium wool, meat
*Distribution:* Worldwide

## 3.3 Breed creation programs implemented in the development of the Meatmaster sheep breed

### 3.3.1 Damara x Ile de France - Breed creation program No.1, executed by FW Peters at Schoongezicht on the Highveld in Gauteng

Damara rams were mated with Ile de France ewes to produce the F1 generation as in figure 3.8. The F1 generation ewes were then mated to F1 generation rams as far as possible unrelated to each other. F2 generations were put to F2 generations of the opposite sex, and then F3 x F3 and F4 x F4 generations followed. The F2 generation was regarded as an appendix A animal, the F3 generation as an appendix B animal and the progeny of the appendix B animals (F4 generation) as a studbook proper animal.

The principle adhered to was to maintain a composition of 50% Damara and 50% Ile de France in the composite and to breed at least four successive generations. Schematically represented if the Damara = A and Ile de France = B.

Then breed A x breed B = AB (50:50) F1
AB x AB = AABB F2 (50:50)
AB x AB = AABB F3 (50:50)
AB x AB = AABB F4 (50:50) = M
with the %RĤV = 50%.

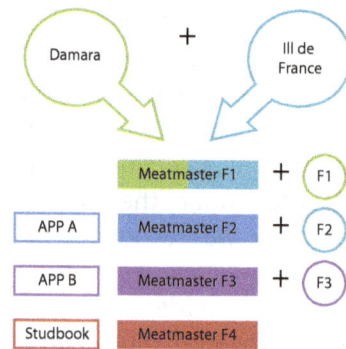

Figure 3.8   Breed creation program No.1

Figure 3.9  Schoongezicht Meatmasters bred by Program No.1 on the Highveld near Meyerton in Gauteng

### 3.3.2 Damara x Meatmaster x Ile de France x Meatmaster - Breed creation program No. 2

Breed creation program No. 1 was followed for a period of nine years where after F3 and F4 Meatmaster rams were mated to Ile de France ewes and Damara ewes to breed new F1 generations in order to increase the genetic diversity. The composition of 50% Damara and 50% Ile de France was however maintained.

F1 rams from Ile de France dams and Meatmaster sires (75% IDF & 25% D) were mated to F1 ewes from Damara mothers and Meatmaster sires (75% D & 25% IDF) during 2005, resulting in a new F2 generation of 50% Damara blood and 50% Ile de France blood. Genetic diversity was increased and this generation was successively mated with Meatmaster rams during 2006 to 2009. If the Damara = A, Ile de France = B and Meatmaster = M, where M = AB, the breed creation program may be represented as follows:

A x M = AM F1 and B x M = BM F1
AM x BM = ABM (25:25:50) = ABAB F2 (50:50) but the % RĤV = 63% in subsequent generations be-

cause this combination now resembles a 3-breed composite of (25% Damara, 25% Ile de France and 50% Meatmaster).

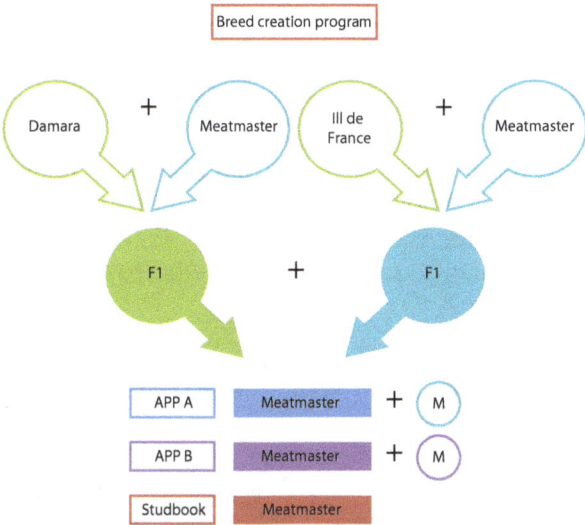

Figure 3.10   Breed creation program No. 2

During 2005 a group of 206 lambs were bred in accordance with this breed creation program. These lambs were subsequently mated with Meatmaster rams bred according to breed creation program no.1 and Meatmaster rams from the Venterstad Meatmaster population bred by CR Collett.

Figure 3.11   Meatmasters bred by Freddie Peters in accordance with Program No.2 on the Highveld

### 3.3.3   Meinfred Meatmasters x Eastern Cape Meatmasters

According to the Meatmaster breed standards the Meinfred Meatmasters had too much wool especially for hotter and dryer areas. The ewes shed their coats reasonably well but the rams maintained a woolen coat for an extended period of time (personal observation).

Stud rams from CR Collett's Meatmaster Stud in the Eastern Cape were introduced since February 2005 to improve the coat covering characteristics of the Meinfred Masters as well as to increase the genetic diversity of the flock.

With the Damara = A, Ile de France = B and White Dorper = C the result of this breed creation program was a Meatmaster comprising of three breeds with unequal representation. The percentage of Damara blood was however maintained at 50%. The result of this breed creation program showed a dramatic improvement in coat covering characteristics and the F1 and F2 generations conformed to Meatmaster breed standards. Three groups of 200 lambs each were bred according to this program during the course of the breed creation program, in 2006 a group of 203 lambs, in 2007 a group of 230 lambs and in 2008 a group of 210 lambs.

AB x AC = AABC (50:25:25) F1
AABC x AABC = AAAABBCC (50:25:25) F2 with the %R$\hat{H}$V = 63%

### 3.3.4   The Damara x Dorper – breed creation program

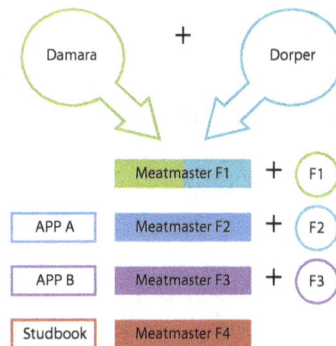

Figure 3.12   The Damara x Dorper breed creation program

Figure 3.13   Meatmasters at La Rochelle in the Venterstad District of the Eastern Cape bred by Clynton Collett

The breed creation program followed by CR Collett in the Eastern Cape and Other Northern Cape breeders such as JP du Plessis, DH Visser and others, who were also former Damara stud breeders, was to implement breed creation programs where Damara ewes were initially mated to White Dorper or Dorper rams (personal communication). Damara = A and Dorper = C:

Then breed A x breed C = AC (50:50) F1
AC x AC = AACC (50:50) F2
AC x AC = AACC F3
AC x AC = AACC F4 = M with %RÂV = 50%
The Hopetown Meatmaster population of Dr Walti Vermeulen was bred from three different breeds, Damara = A, Dorper = C and Van Rooy = D (personal communication).
A x C = AC and C x D = CD
Then AC x CD = ACCD (25:50:25) F1
ACCD x ACCD = ACCD F2
ACCD x ACCD = ACCD F3 with %RÂV = 63%

Figure 3.14  Hopetown Meatmasters bred by Dr Walti Vermeulen

The Prieska Meatmaster population bred by Ms CM du Toit also comprises a component from three parent breeds, Damara = A, Dorper = C and SA Mutton Merino = E (personal communication).
A x C = AC and A x E = AE
Then AC x AE = AACE (50:25:25) F1 with %RÂV = 63%
and AACC x AACE = AAAACCCE (50:38:12) F2. If this combination is retained in subsequent generations the %RÂV = 59%.

Figure 3.15  Prieska Meatmasters bred by Christine Du Toit

Dr JJ Steyn in the Bloemfontein area obtained special permission to import semen of the Wiltshire Horn sheep breed. The Wiltshire Horn breed has good general conformation and a natural characteristic to shed its woolen coat in summer. Strong horns and a shorter breeding season may be a disadvantage. JJ Steyn was of the opinion that inclusion of a 25% Wiltshire Horn component in a Meatmaster flock would be a good combination. The breed creation program then consisted of three breeds (Steyn, 2006).
A = Damara, C = Dorper and F = Wiltshire Horn (Steyn, 2006).

F1 Damara x Wiltshire Horn original 1997 ewes

Figure 3.16    F1 Damara x Wiltshire Horn ewes bred in Australia

A x C = AC and A x F = AF
Then AC x AF = AACF (50:25:25) F1
AACF x AACF = AACF F2
AACF x AACF = AACF F3
For this combination the %RÂV = 63%.
Another possibility is to include 4 parent breeds where A = Damara, C = Dorper,
D = Van Rooy and F = Wiltshire Horn, resulting in:
A x C = AC and D x F = DF
Then AC x DF = ACDF (25:25:25:25) F1
ACDF x ACDF = ACDF (25:25:25:25) F2
For this breed combination the %RÂV will increase to 75%.

### 3.3.5  The Gellapper breed creation program in Namibia

Damara and Dorper sheep were used in the following breed creation program by the Gellap-Ost research station at Keetmanshoop in Namibia (personal communication) resulting in the Gellapper with about 70% Damara and 30% Dorper genes. The back crossing procedure will have the effect of gradually decreasing the %RÂV but for the final composition the %RÂV = 42%.

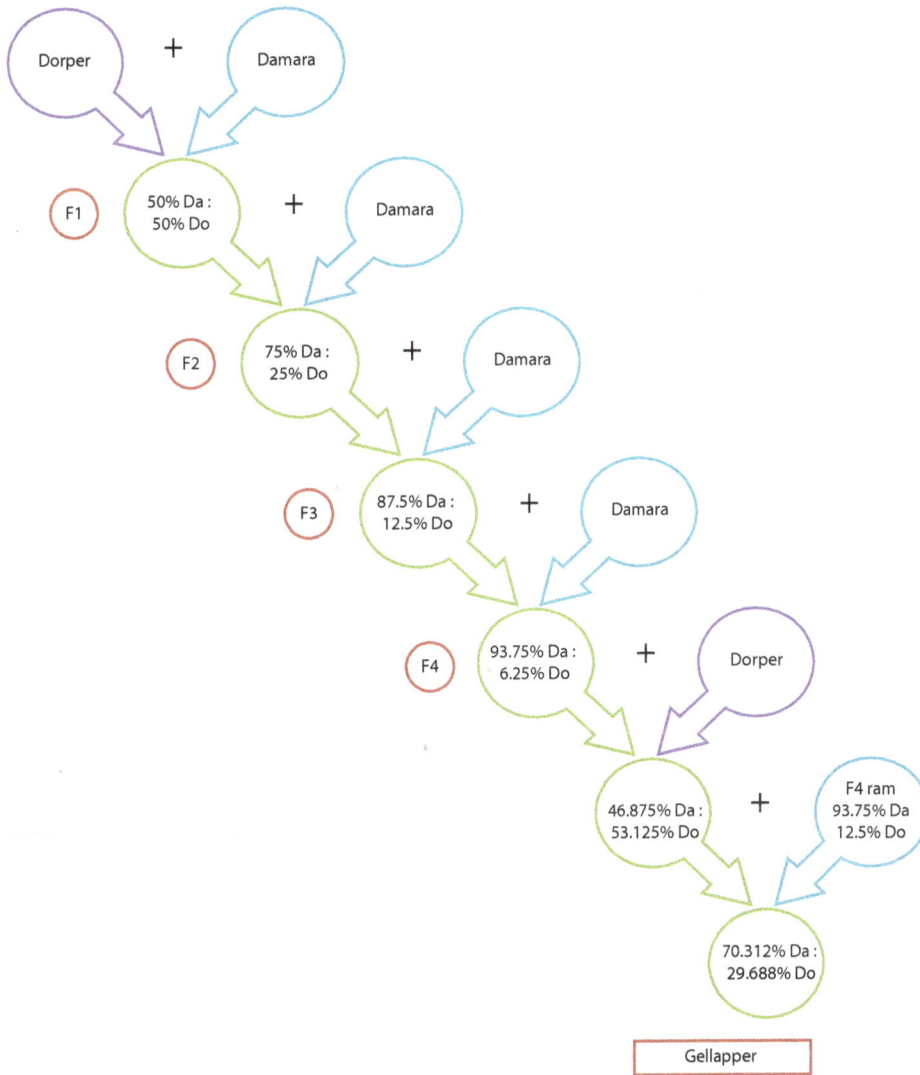

*Figure 3.17 Gellapper breed creation program in Namibia*

*Figure 3.18   Gellapper Meatmasters bred by Gellap-Ost research station in Namibia*

### 3.3.6  The Damper, Meatmaster and Drought-master breed creation programs in Australia

Russel (2010) reported on the Genelink Meatmaster sheep in South Australia. Damara rams were mated to Wiltshire horn, Wiltshire x Suffolk and Suffolk ewes in 1997, 1998, 1999 and the resulting lambs bred back to the Damara (F2 and F3 generations) where after several other crossbreeds have been used including Van Rooy x Wiltshire Horn. White Dorper rams were introduced to improve conformation.

Selection for the coat shedding trait was applied to obtain the clean shedding phenotype, as meat production was the main aim.

During 2006 the Red Droughtmaster ewe type was isolated and mated to rams with red genetics. The Australian Meatmaster is able to produce a good quality and weight of lamb in 4-5 months (45kg) on good grazing and carcasses are acceptable to the slaughter market.

The Meatmaster is marketed under different names in Australia, Droughtmaster is a red variation and Damper is a synonym for Meatmasters in Western Australia.

There is good export potential for Meatmaster breeding sheep into the Arab countries, Malaysia, Indonesia and China.

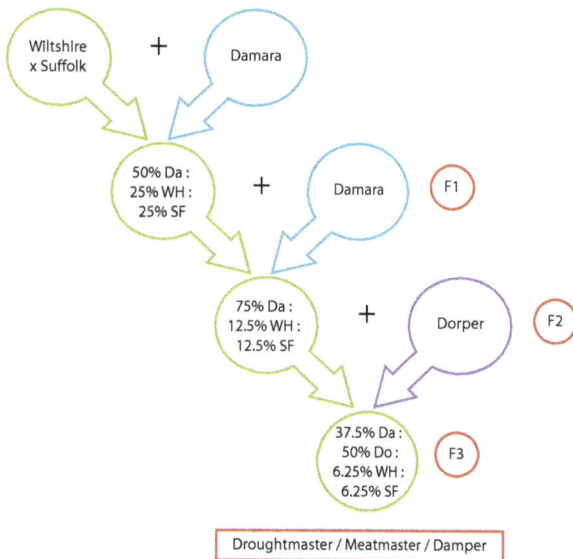

Figure 3.20  F3 Damara x Suffolks incorporated in the Genelink Meatmasters in Australia

Figure 3.19  Droughtmaster breed creation program in Australia

If the four-breed composition as in fig 3.19 is retained in subsequent generations the expected %RĤV = 60%.

Several other crossbreeds were attempted including Van Rooy x Wiltshire Horn, Damara x Van Rooy, Namaqua Afrikaner and East Friesian crosses (Russel, 2010). Since 1995 many indigenous sheep breeds from South Africa were imported by breeders in Australia. It is therefore relatively easy and practical for breeders in Australia to follow the Meatmaster breed creation programs developed in South Africa. The Meatmaster Sheep Breeders' Society of South Africa is prepared to assist breeders internationally to preserve the true Meatmaster genetics and to accept membership from breeders in foreign countries in order to establish the Meatmaster breed worldwide in an organised way so that proper recording will be possible to preserve Meatmaster breed standards.

Figure 3.21  Genelink Meatmasters bred by Dennis Russel in Australia

### 3.3.7 Recommended breed creation programs

Personal observation by both FW Peters and CR Collett confirmed that combinations where the Damara component (A) is less than 50% lose some of the most valuable characteristics of the Meatmaster sheep breed. The composition by which the resulting 50% is made up is less important in terms of the prescribed breed standards and may vary in accordance with local environmental conditions and the requirements of different ecosystems.

It is recommended that the Meatmaster should have 50% Damara genes, a Dorper component and may include varying components from Ile de France, Van Rooy, SA Mutton Merino, Dormer, Wiltshire Horn and other sheep breeds. The guideline that should be adhered to is to meet the Meatmaster breed standards. Some of the Damara component could be replaced by Pedi sheep for example due to the close relationship in the characteristics of the Damara and Pedi sheep breeds. The inclusion of some Pedi sheep genes in the final composition may be advantageous to the adaptation and establishment of Meatmaster sheep in the heartwater areas of the North West, Limpopo and Mpumalanga Provinces where Pedi sheep are indigenous and well adapted.

### 3.3.7.1 Other breeds that can be evaluated in cross breeding programs

In the Meatmaster Newsletter of January 2006, Dr JJ Steyn recommended the following breeds for possible inclusion in the gene pool of the Meatmaster as a developing breed - The Damara, Dorper/White Dorper and Van Rooy as successfully proven combinations and for possible mother lines the Blackhead and Speckled Persian breeds, Ronderib Afrikaner and Namaqua Afrikaner breeds.

For grain producing areas Steyn recommends the inclusion of dual purpose sheep breeds like the Ile de France, Dormer, SA Mutton Merino and the recently imported Wiltshire Horn exotic breed with its unique coat cover shedding characteristics.

### 3.3.7.2 Breeding procedures recommended

For the breeding of a hair type non-fat tailed meat sheep breed like the Meatmaster, CR Collett (2003) recommended the following breeding procedures:

- Use a Damara ram on Dorper ewes for one generation and there after Meatmaster rams on the progeny.
- Use Damara rams on SA Mutton Merino ewes for two generations and there after a Dorper ram on the second generation followed by using Meatmaster rams on the progeny.
- Use Meatmaster rams on Van Rooy and Persian type of ewes for successive generations.
- Use either White Dorper or Dorper rams on Damara ewes for one generation and there after Meatmaster rams on the progeny.

### 3.3.8 Recommended selection objectives

- Fertile sheep, good mothers that are able to rear their lambs in veld conditions without assistance.
- Rams with a good serving ability.
- Animals that can walk long distances

- Lambs with a slaughter weight of between 12kg – 25kg and an average carcass grading of A3.
- *"Easy care"* sheep with no tail docking, no shearing, minimal inoculation, dosing and dipping.

### 3.3.9 Expected final genetic composition of the Meatmaster sheep

The final genetic composition of the Meatmaster sheep could be envisaged to be somewhat like the following:

AAAAABCCCDEF = M (50:7.5:20:7.5:7.5:7.5) with a %R$\hat{H}$V = 69% varying to AAAAAAAABCCCCCDDEF = M (40:05:25:20:05:05) with a %R$\hat{H}$V = 73% and where A = Damara, B = Ile de France, C = Dorper, D = Van Rooy, E = SAMM and F = Wiltshire Horn.

A modern composite sheep breed with a high percentage of retained hybrid vigor and breed complementarity. The optimum genetic composition will be ecosystem dependent and the Meatmaster Breeders' Society recognised that there may be a degree of difference in type necessitated by different environmental conditions.

The wide genetic base and open ended breed creation programs followed by Meatmaster breeders will ensure high levels of heterozygosity and that the best characteristics from a large number of sheep breeds will be combined in the composite Meatmaster sheep breed.

## 3.4 Exotic Breeds which could add future value to the Meatmaster sheep

The following exotic hair type sheep could possibly add value to the Meatmaster sheep breed if they could be introduced via the breeding of F1 generations from Damara sheep x Exotic Breed, which can then be bred with selected Meatmasters.

These breeds are the Barbados Blackbelly, which could improve the prolificacy of Meatmasters. The Katahdin, which has good conformation characteristics and the Red Maasai sheep, which is renowned for its resistance to gastro-intestinal nematode parasites (Baker et.al, 1999). Schoenian

(2009) provides short summaries of their most important characteristics, breed categories and breed distribution.

### 3.4.1 Barbados Blackbelly

The Barbados Blackbelly is an indigenous breed to the Caribbean island of Barbados. It descends from sheep brought to the islands from West Africa during the slave era. Blackbellies are *"antelope like"* in appearance, brown tan or yellow in colour, with black points and under parts. Both ewes and rams are polled or have only small scurs or diminutive horns.

They may have some visible fuzzy wool undercoat within their hair coat, but it should shed along with the hair each year. Barbados Blackbellies are noted for their extreme hardiness and reproductive efficiency. They are one of the most prolific sheep breeds in the world (Schoenian, 2009). White Dorper rams are already used on Barbados Black-belly ewes in Texas in a hair sheep development program (Walker, 2010).

*Breed category:* hair (meat)
*Breed distribution:* Caribbean, Mexico, South America

### 3.4.2 Katahdin

The Katahdin is an improved breed of hair sheep, the first hair breed to meet North American industry standards for carcass quality. The Katahdin is a cross between British meat breeds, notably the Suffolk, African Hair sheep, specifically the St. Croix, and later the Wiltshire Horn. They were developed in the 1950's by amateur geneticist Michael Piel and take their name from Mt. Katahdin in Maine where the Piel farm was located.

The Katahdin is an easy-care, low-maintenance meat-type sheep that is naturally tolerant of climatic extremes and capable of high performance in a variety of environments. One of the most outstanding characteristics of the Katahdin is its natural resistance to internal parasites. The Katahdin is one of the most popular breeds of registered sheep in the U.S. (Schoenian, 2009).

*Breed categories:* hair (meat)
*Distribution:* North America, Caribbean, Asia

### 3.4.3 Red Maasai

The Red Maasai is an East African fat-tailed type of hair sheep used for meat production. They are found in northern Tanzania, south central Kenya, and Uganda. Red Maasai are red-brown, occasionally pied. Males are horned or polled. Females are usually polled. Red Maasai are known for being resistant to internal parasites (Baker et al., 1999).

*Breed category:* hair (meat), fat-tailed
*Distribution:* Africa, Europe

## 3.5 Hair sheep breed creation research and development in the USA and UK

Walker (2010) reported in his scholarship report: *"The United States Meat Animal Research Center (USMARC) is a US Department of Agriculture (USDA) facility administered by the Agricultural Research Service (ARS) comprising 35,000 acres in Nebraska. There are female breeding populations of 6,500 cattle of 18 breeds and 3,000 sheep of 10 breeds. Scientists at the facility develop new technology to increase the efficiency of livestock production. A current project is the development of an easy-care maternal line of hair sheep that can raise triplets on pasture without labour or supplementary feed.*

*There is further ongoing research evaluating the ability of the Dorper, White Dorper, Katahdin, Dorset and Rambouillet through comparison of crossbred females in both high and low input systems of production. Although having lower prolificacy than other crosses, the Romanov X White Dorper ewes averaged 2.04 lambs reared per ewe lambing. Some 62% of triplet bearing ewes weaned their entire litters from pasture alone with no shepherding input, shattering the common perception of many sheep producers and scientists.*

*The Romanov x White Dorper was the best performer and hence comprises 75% of the mix. The inclusion of Katahdin and USMARC Composite is partly for easy-care traits (parasite tolerance and hair of Katahdin) and marketing (size, conformation and scrapie resistance of USMARC Composite), but also to increase the retention of hybrid vigour effects from 50% in a two way cross to 66.6% in the four way cross. This breeding program is already under way and 2010 will see the first matings of F1 ewes to F1 rams with the goal of ultimately reaching 4,000 ewes. The F2 generation will then be suitable for evaluation with other maternal breeds in a low input pasture lambing system.*"

*Figure 3.22  Romanov*

*Figure 3.23  Katahdin*

Another international example is the Easycare breed developed by Lolo Owen in Wales. The aim was to produce a breed of sheep which would require minimal shepherding and veterinary care and yet offer good meat yields and lambing ratios. This breed sheds its wool, reducing or eliminating labour costs associated with shearing and dagging lambs prior to slaughter (Collins and Conington, 2010).

## 3.6   Conclusions

From an African perspective it is recommended that further trials be done with breed creation programs which combine the Meatmaster and Red Maasai sheep in similar proportions to trials previously done with Dorper and Red Maasai sheep in tropical and sub-tropical areas, to determine genetic resistance to gastro-intestinal nematode parasites.

There is a worldwide need for an easy-care, clean skin hair type, mutton producing sheep. Current research in the USA also emphasise this need. Similar projects are investigated in the UK.

The Damara sheep is the parent breed which contributes the most to the desired characteristics and Meatmaster breed standards can only be achieved if a sufficient percentage of Damara genes are present in the final product. As soon as the genetic contribution of the Damara parent breed is diluted to less than 50%, valuable traits of the Meatmaster breed are lost as the focus shifts to improved conformation, resulting in heavier ewes and a consequent decrease in productivity.

The maintenance of the required genetic balance will be of great importance for Meatmaster breeders in order to conserve the future agro-economic value of the breed.

# Klaas Grobler
# Kern Meatmasters

Douglas

Klaas & Sonia Grobler  084 790 8191
kernmeatmasters@gmail.com

# Chapter 4  Genetic profile of the locally developed Meatmaster sheep breed in South Africa based on micro-satellite analysis

## 4.1  Introduction

Breeds of domesticated animals that are selected for optimal survival and production under specific local conditions provide considerable benefits to breeders. Indigenous and locally developed sheep breeds are therefore an important asset because of the unique combinations of adaptive traits that developed to respond effectively to the pressures of the local environment (Buduram, 2004). These adaptive traits include tolerance to various diseases, fluctuations in feed quality, extreme climatic conditions and the ability to survive and reproduce for long periods of time (Hammond, 2000). Many examples of locally adapted sheep breeds have been reported, including breeds such as the Muzzafarnagri from India (Arora and Bhatia, 2004) and the Chiapas from Mexico (Quiroz et al., 2008). Well-known examples of composite breeds developed in Southern Africa are the Dorper and Van Rooy (Ramsey et al., 2000).

Genetic aspects of the development of the new South African Meatmaster sheep breed, founded through the combination of the well adapted indigenous Damara sheep breed with other locally developed breeds, as well as exotic sheep breeds are discussed. The Damara indigenous breed was predominantly included in the genetic make-up of the Meatmaster as the maternal ancestor. This breed contributed specifically to the characteristics of colour variation, coat shedding ability and adaptability. The contribution of the Ile de France, Dorper and Mutton Merino breeds was primarily aimed at the improvement of conformation and carcass quality. The Van Rooy breed was introduced to improve coat covering characteristics. Specific selection guidelines that were encouraged for the Meatmaster breed were aimed at protection against harsh climatic conditions, enhancement of meat production, improved carrying capacity, mobility in various habitats, carcass quality that meet market standards and suitable behaviour (strong herd instinct).

The resultant Meatmaster sheep breed is a composite landrace breed, utilizing a large gene pool from which base selection to achieve a prescribed set of breed standards, based on agro-economic

principles, could be done. The Meatmaster breed was formally declared a developing breed by the National Department of Agriculture in South Africa in 2007 in the Government Gazette (nr. 29898 of 2007).

Breeding programs, phenotypic descriptions and examples of the breed, performance, breed standards and agro-economic considerations are described in chapters included in this book. The commendable approach of the Meatmaster Sheep Breeders' Society of South Africa, not to limit the gene pool to prescribed genetic combinations in the development and establishment of the Meatmaster sheep breed, created the potential of a very wide genetic variation which is essential for the long term viability of the breed.

*Figure 4.1    Map indicating the localities where blood samples were taken*

## 4.2    Objectives of the DNA analysis of Meatmaster sheep

The objectives of the DNA analysis of the Meatmaster were:

1. to determine whether the Meatmaster forms a recognizable genetic group,
2. to describe the pattern and scale of genetic divergence between the Meatmaster and its parent breeds,
3. to describe relationships among individual

**41**

Meatmaster populations and

4. to describe within-population diversity in Meatmaster populations.

The genetic profile of the Meatmaster at its onset provides a valuable reference point against which possible future genetic drift can be measured.

## 4.3 Procedures followed to perform the Meatmaster DNA analysis

### 4.3.1 Meatmaster populations sampled and sample collection (2006)

Blood samples were taken by FW Peters from the following Meatmaster (MM=pooled samples for the breed) populations: 55 individuals from Meyerton (Gauteng Province, abbreviation: MMe) and 40 each from Venterstad (Eastern Cape, MVe), Hopetown (Northern Cape, MHo) and Prieska (Northern Cape, MPr). Each of these populations was developed from a slightly different genetic base, with different proportions of breeds used for breed formation. However, all populations are characterized by a significant component of Damara (Dam) genetic material among the ancestors (25-75%). These breeding programmes commenced between the late 1980s and early 1990s. Current generation animals were sampled for this project. Samples were also taken from 40 Ile de France sheep (IFr), a parent breed of the Meatmaster.

Table 4.1   Sheep blood samples collected by FW Peters

| Breed/population | No. of samples | Breeder | Location |
|---|---|---|---|
| Meatmaster (MMe) | 55 | FW Peters | Meyerton, GP |
| Ile de France (IFr) | 12 | PJ de Bruyn | Meyerton, GP |
| Ile de France (IFr) | 28 | W Rossouw | Sannieshof, NW |
| Meatmaster (MVe) | 40 | CR Collett | Venterstad, EC |
| Meatmaster (MHo) | 40 | PvdW Vermeulen | Hopetown, NC |
| Meatmaster (MPr) | 40 | CM du Toit | Prieska, NC |

### 4.3.2 DNA extraction from Meatmaster sheep blood samples

Permission was obtained to make use of the existing DNA database of reference populations at the Agricultural Research Council (ARC, Irene) for other parent breeds (SA Mutton Merino (SAM), Damara (Dam), Dorper (Dor) and Van Rooy (VaR)).

The DNA database of the Ronderib Afrikaner breed (RoR) from the ARC was also included, since this breed contributed significantly to the Van Rooy breed. Genotypes of 10 unrelated males and 30 unrelated females from each of the parent breeds were used. Genotypes of 34 individuals of the Namaqua Afrikaner (NaA) from the DNA database at the ARC were used as an unrelated group because this breed is completely unrelated to the Meatmaster or any of its parent breeds.

The advent of DNA technology has revolutionised the identification and genetic characterization of farm animals. Each animal has a unique DNA profile that distinguishes it from any other even its siblings, but even more important is the fact that each profile is a combination of the two parental profiles (Buduram, 2004). DNA was extracted from whole blood samples using the Wizard Genomic DNA Purification kit (Miller et al., 1998).

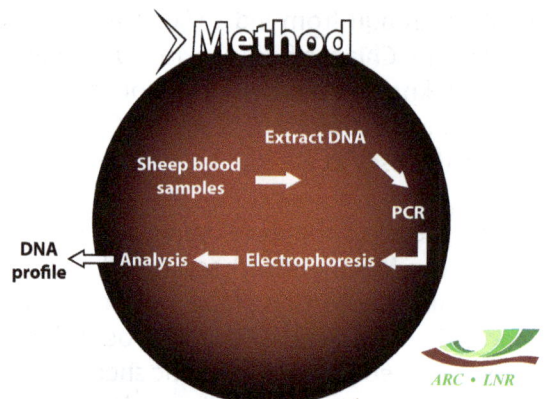

Figure 4.2   Schematic illustration of the method applied to obtain the DNA profile

### 4.3.3 Micro-satellite markers selected

DNA micro-satellite sequences are valuable genetic markers due to their dense distribution in the genome and are relatively easy to detect. Micro-satellites offer several advantages. They are relatively easy to isolate in different species, different loci can be used according to the level of variation, ranging from very low to extremely high (Beaumont and Bruford, 1999), they can be easily amplified by PCR and thus used on a wide range of sample material such as blood, hair, meat and skin, and their genetic systems are easily automated enabling the analysis of a large number of samples (Luikart et al., 1999). The resolving power of micro-satellites contributes to its popularity amongst researchers.

*Table 4.2   Characteristics of selected micro-satellites*

| Micro-satellite marker | Forward Primer Sequence | Reverse Primer Sequence | Size Range | References |
|---|---|---|---|---|
| INRA63 | ATTTGCACAAGCTAAATCTAACC | AAACCACAGAAATGCTTGGAAG | 165-225 | Vaiman & Mercier (1994) |
| TGLA53 | CAGCAGACAGCTGCAAGAGTTAGC | CTTTCAGAAATAGTTTGCATTCATGCAG | 130-175 | Crawford et al.   (1995) |
| CSSM36 | GGATAACTCAACCACACGTCTCTG | AAGAAGTACTGGTTGCCAATCGTG | 150-210 | Kemp et al. (1993) |
| MGTG4B | GAGCAGCTTCTTTCTTCTCATCTT | GCTCTTGGAAGCTTATTGTATAAAG | 120-145 | Steffen & Eggen (1993) |
| OARFCB20 | AAATGTGTTTAAGATTCCATACAGTG | GGAAAACCCCCATATATACCTATAC | 60-120 | Buchanan & Crawford (1992) |
| ETH225 | GATCACCTTGCCACTATTTCCT | ACATGACAGCCAGCTGCTACT | 120-160 | Steffen & Eggen (1993) |
| TGLA57 | GCTTTTTAATCCTCAGCTTGCTG | GCTTCCAAAACTTTACAATATGTAT | 80-120 | Steffen & Eggen (1993) |
| CSRD247 | GGACTTGCCAGAACTCTGCAAT | CACTGTGGTTTGTATTAGTCAGG | 220-260 | Kemp et al. (1993) |
| ILSTS087 | AGCAGACATGATGACTCAGC | CTGCCTCTTTTCTTGAGAGC | 130-175 | Kemp et al. (1993) |
| RM004 | CAGCAAAATATCAGCAAACCT | CCACCTGGAAGGCCTTTA | 100-160 | Kossarek & Grosse (1993) |

Microsatellite markers were used for genetic characterization in this project because this technique has been proven useful for the analysis of genetic differentiation among sheep populations (Saitbekova et al., 2001, Tomasco et al., 2002, Quiroz et al., 2008). An ovine microsatellite marker set that would enable comparison with the existing ARC database was compiled as described in Table 4.2.

### 4.3.4   Polymerase Chain Reaction (PCR) and Genotyping of Meatmaster DNA

The Polymerase Chain Reaction (PCR) amplification was performed using a Perkin Elmer Gene Amp PCR System 9700 thermocycler. Genotyping was carried out on an automated ABI 377 DNA sequencer (Perkin Elmer, Foster City, USA) with fragments separated using a 5% Long Ranger/6M gel.

### 4.3.5   Statistical analysis of Meatmaster DNA Data

Statistical analysis of the Meatmaster DNA data were performed at the University of the Free State in collaboration with Professor JP Grobler. The GENECLASS programme was used for individual specific analysis. Allele frequencies within breeds and populations were calculated using MSToolkit (Park, 2001). MSToolkit was also used to prepare or initiate input files for all other software used.

A Bayesian-based assignment test (Pritchard et al., 2000) of multilocus genotypes was used to identify the true number of populations (clusters) and assign individuals probabilistically to each cluster (using STRUCTURE software - Falush et al., 2003). Each of the 419 sheep genotyped was entered as an individual with putative population of origin. The model used was based on an assumption of admixed ancestry and correlated allele frequencies.

Genetic divergence among populations was quantified using frequency-based measures based on the infinite alleles model (IAM) and the stepwise mutation model (SMM) respectively. As representative of the IAM, we calculated $F_{ST}$ values (Wright, 1965) between all population pairs using ARLEQUIN software (Excoffier et al., 2005). The SMM was accommodated by calculating pairwise $R_{ST}$ values (Slatkin, 1995) using RST-CALC software (Goodman, 1997). Calculations of $F_{ST}$ and $R_{ST}$ were done with (i) all Meatmaster individuals pooled as one population (MM), and (ii) with each of the four Meatmaster populations included as a separate subunit. Genetic distances among populations were determined using Nei's standard genetic distance (Nei, 1972), as implemented in DISPAN software (Ota, 1993). Genetic distances were then used to construct phylogenetic trees using the unweighted pair group-method with arithmetic mean (UPGMA) method (Sneath and Sokal, 1973), with 1000 bootstrap replications.

*Table 4.3   Proportion of membership of 11 breeds and populations of sheep to 10 nominal clusters, based on Bayesian analysis. Values blocked with double lines indicate clusters dominated by single established parent breeds, values blocked with single lines indicate three clusters containing predominantly Meatmaster individuals*

| | **Cluster:** | | | | | | | | | |
| | 1 | 2 | 3 | 4 | 5 | 6 | 7 | 8 | 9 | 10 |
|---|---|---|---|---|---|---|---|---|---|---|
| SAM | 0.01 | 0.012 | **0.869** | 0.030 | 0.008 | 0.012 | 0.026 | 0.011 | 0.010 | 0.012 |
| Dam | 0.024 | 0.018 | 0.011 | 0.036 | 0.020 | 0.030 | 0.01 | **0.792** | 0.034 | 0.024 |
| Dor | 0.029 | 0.018 | 0.012 | 0.031 | **0.830** | 0.017 | 0.014 | 0.011 | 0.018 | 0.018 |
| VaR | **0.849** | 0.021 | 0.012 | 0.014 | 0.018 | 0.014 | 0.033 | 0.017 | 0.01 | 0.012 |
| IFr | 0.011 | 0.037 | 0.031 | 0.045 | 0.015 | 0.028 | **0.787** | 0.013 | 0.017 | 0.015 |
| RoR | 0.011 | **0.929** | 0.007 | 0.008 | 0.006 | 0.006 | 0.005 | 0.007 | 0.011 | 0.009 |
| Mme | 0.032 | 0.027 | 0.036 | **0.150** | 0.040 | **0.443** | 0.109 | 0.065 | **0.081** | 0.016 |
| MVe | 0.025 | 0.014 | 0.010 | **0.113** | 0.026 | **0.228** | 0.014 | 0.091 | **0.466** | 0.013 |
| Mho | 0.084 | 0.017 | 0.025 | **0.244** | 0.102 | **0.082** | 0.032 | 0.074 | **0.314** | 0.026 |
| MPr | 0.027 | 0.017 | 0.046 | **0.277** | 0.074 | **0.094** | 0.031 | 0.118 | **0.293** | 0.021 |
| NaA | 0.036 | 0.016 | 0.012 | 0.016 | 0.021 | 0.027 | 0.010 | 0.015 | 0.016 | **0.831** |

A hierarchical analysis of the distribution of total genetic diversity in all the individuals screened was done through analysis of molecular variance (AMOVA; Michalakis and Excoffier, 1996), using ARLEQUIN software. For this analysis, two population structures were used. (i) Firstly, all established sheep breeds were entered as individual groups. Meatmaster populations were added as an additional group, with the overall Meatmaster group further partitioned into four separate populations. (ii) To allow for possible inflation of the among-group variance component due to the presence of the out-group (Namaqua Afrikaner), the analysis was repeated without this breed.

To quantify levels of diversity within individual breeds and populations, observed heterozygosity (Ho), Nei's unbiased heterozygosity (Hz; Nei, 1987) and average number of alleles per locus (A) were calculated using MSToolkit.

## 4.4 Results of Meatmaster DNA analysis

Scrutiny of allele frequencies for breed-specific alleles showed that 104 out of 120 alleles scored were shared by two or more breeds. Among the 16 alleles unique to individual breeds, 14 occurred at frequencies below 0.05 (0.013-0.038). Possibly breed-specific alleles at frequencies higher than 0.05 were found only at locus INRA63 in Namaqua Afrikaner (0.162) and CSRD247 in the Ile de France (0.063). In Meatmaster populations, unique alleles occurred at low frequencies only, with one allele at locus MGTG4B (frequency=0.013) and three at OARFCB20 (frequencies=0.013-0.038).

Calculation of the posterior probabilities of K showed the highest probability for a real structure consisting of 10 populations (Table 4.3). All the established breeds were well defined, with 78.7-92.9% of individuals for each breed assigned to distinct breed-specific clusters. The majority of Meatmaster individuals (67.4% of MMe; 80.7% of MVe; 64.0% of MHo and 66.4% of MPr) were spread over the remaining three clusters (Clusters 4, 6 and 9, blocked with single lines in Table 4.3, containing only limited numbers of sheep from the established breeds (1.0%-4.5%). Bar plots showing the proportion of membership of each individual to one or more of the 10 real clusters identified is presented in Figure 4.4.

Values for coefficients of genetic differentiation corresponded well to known relationships between breeds, but with a slightly better fit obtained for the SMM-based $R_{ST}$ values compared to $F_{ST}$ values. Values of $R_{ST}$ between MM and its parent breeds ranged from 0.081-0.162, whereas $R_{ST}$ between the Meatmaster and the unrelated group was much higher at 0.266 (Table 4.4). $F_{ST}$ values were 0.049-0.156, and 0.085 for the above-mentioned groups respectively. Treatment of the four Meatmaster populations as separate groups (Table 4.5) also resulted in $R_{ST}$ values that corresponded well with known relationships. Pair-wise comparisons among Meatmaster populations yielded $R_{ST}$ values of 0.004-0.028 (Table 4.5). By comparison, $R_{ST}$ values between combinations of Meatmaster and parent breeds were higher at 0.051-0.194, and still higher between Meatmaster populations and the out-group (0.237-0.309). Values of $F_{ST}$ among Meatmaster populations were 0.012-0.033, with slightly higher values (0.047-0.194) among Meatmaster populations and parent breeds. $F_{ST}$ values of 0.093-0.120 were obtained between Meatmaster populations and the out-group (Table 4.5). Differentiation between all pair-wise combinations of populations is significant at the 95% level.

Genetic distances between all population pairs are presented in Table 4.6 and Figure 4.5. Genetic distances among Meatmaster populations (0.022-0.079) were smaller than distances between Meatmaster and established breeds (0.142-0.437) or distances among established breeds (0.224-0.694). The dendrogram shows strong bootstrap support (97%) for a distinct cluster containing all the Meatmaster populations, with no other breeds included in the cluster.

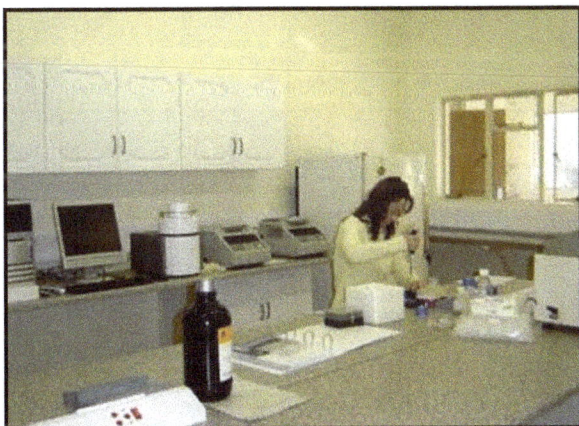

Fig 4.3   *The ARC animal genetics Lab where Meatmaster DNA analysis was done*

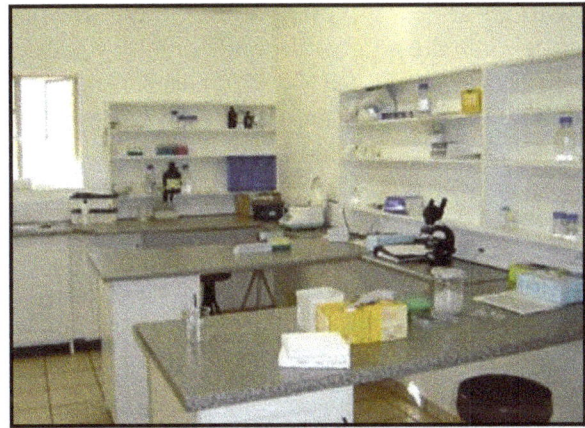

Figure 4.4   *Bar plot showing individual sheep by breed or population. Coloured zones on each vertical bar show the proportion of membership of an individual to each of one or more of 10 real clusters identified.*

Unbiased heterozygosity, observed heterozygosity and average number of alleles per locus for all 11 sheep populations and pooled MM are listed in Table 6.7 The average heterozygosity in Meatmaster populations (Hz=0.634-0.705) were comparable to the higher end of the range in the parent breeds (0.524-0.705). Similarly, Ho values (0.519-0.616 compared to 0.362-0.631) and A in

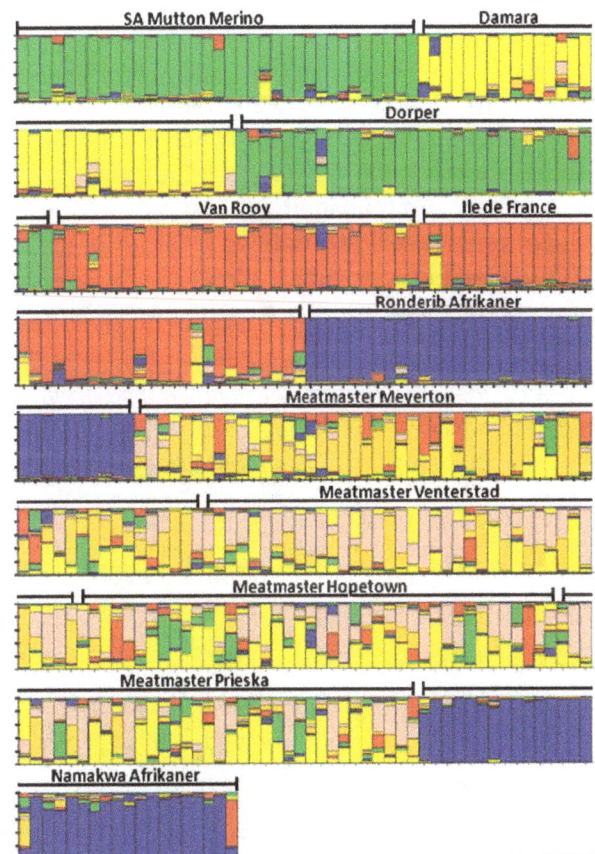

Meatmaster populations (6.7-8.2) were mostly higher than the number in parent breeds (4.3-6.9). Values of Hz (0.689), Ho (0.570) and A (9.9) were higher in MM than in any other breed. Values for Hz (0.592), Ho (0.397) and A (4.90) in the out-group (NaA) were relatively low compared to values for the Meatmaster and its parent breeds.

## 4.5  Discussion of Meatmaster DNA statistics

The FAO defines a breed as a specific group of domestic animals with definable and identifiable external characteristics that differentiate it from other similar groups of the same species (Quiroz et al., 2008). The classification of sheep breeds based on morphological traits alone is not always consistent with historical and/or genetic data (Rendo et al., 2004) and discriminating genetic characterization of farm animals is necessary. It is therefore important for the future recognition and management of the developing Meatmaster breed to establish whether genetic differences between populations of this breed, and parent breeds, indicate real genetic distinc-

tiveness or simply reflect random genetic drift. The frequency- and multilocus-based measures used provide considerable genetic support for the recognition of the Meatmaster as a distinct breed. This is in contrast to the absence of unique, breed specific alleles. Only four alleles unique to the Meatmaster were found, and these occurred at very low frequencies of 0.013-038. However the established ancestral breeds analysed are characterized by a similar lack of unique alleles with the highest frequency of a breed-specific allele of 0.162 in one breed. These values are too low to be diagnostic for breeds, and the characterization of breeds, both those already established, as well as the Meatmaster, is thus dependent on frequency- and multilocus-based techniques.

Bayesian-based assignment tests based on multilocus genotypes can add valuable information on the distinctiveness of animal populations, and was applied for this purpose by Quiroz et al. (2008) in Mexican sheep populations. In this evaluation, the assignment test firstly confirmed that the parent breeds, as well as the out-group, are well defined genetic units.

*Table 4.4  Genetic differentiation between eight sheep breeds, with R$_{ST}$ values below the diagonal and F$_{ST}$ values above the diagonal. Value ranges blocked with single lines indicate drift between Meatmaster and parent breeds, values blocked with double lines show drift between MM and the out-group*

|      | SAM   | Dam   | Dor   | VaR   | IFr   | RoR   | MM    | NaA   |
|------|-------|-------|-------|-------|-------|-------|-------|-------|
| SAM  | -     | 0.158 | 0.206 | 0.175 | 0.091 | 0.213 | 0.115 | 0.158 |
| Dam  | 0.159 | -     | 0.133 | 0.104 | 0.167 | 0.190 | 0.049 | 0.132 |
| Dor  | 0.180 | 0.119 | -     | 0.119 | 0.169 | 0.236 | 0.068 | 0.142 |
| VaR  | 0.163 | 0.035 | 0.083 | -     | 0.159 | 0.204 | 0.064 | 0.163 |
| IFr  | 0.100 | 0.222 | 0.154 | 0.177 | -     | 0.242 | 0.096 | 0.151 |
| RoR  | 0.127 | 0.177 | 0.293 | 0.203 | 0.242 | -     | 0.156 | 0.205 |
| MM   | 0.081 | 0.103 | 0.162 | 0.107 | 0.087 | 0.116 | -     | 0.085 |
| NaA  | 0.271 | 0.105 | 0.217 | 0.059 | 0.368 | 0.284 | 0.266 | -     |

*Table 4.5   Genetic differentiation between 11 sheep breeds/populations, with R$_{ST}$ values below the diagonal and F$_{ST}$ values above the diagonal. Value ranges blocked with single lines indicate drift between specific Meatmaster populations and parent breeds, values blocked with double lines show drift between Meatmaster populations and the out-group*

| | SAM | Dam | Dor | VaR | IFr | RoR | Mme | MVe | MHo | MPr | NaA |
|---|---|---|---|---|---|---|---|---|---|---|---|
| SAM | - | 0.158 | 0.206 | 0.175 | 0.091 | 0.213 | 0.111 | 0.151 | 0.138 | 0.098 | 0.158 |
| Dam | 0.159 | - | 0.133 | 0.104 | 0.167 | 0.190 | 0.047 | 0.066 | 0.069 | 0.065 | 0.132 |
| Dor | 0.180 | 0.119 | - | 0.119 | 0.169 | 0.236 | 0.086 | 0.092 | 0.062 | 0.094 | 0.142 |
| VaR | 0.163 | 0.035 | 0.083 | - | 0.159 | 0.204 | 0.071 | 0.091 | 0.067 | 0.092 | 0.163 |
| IFr | 0.100 | 0.222 | 0.154 | 0.177 | - | 0.242 | 0.083 | 0.133 | 0.110 | 0.096 | 0.151 |
| RoR | 0.127 | 0.177 | 0.293 | 0.203 | 0.242 | - | 0.155 | 0.189 | 0.194 | 0.175 | 0.205 |
| MMe | 0.051 | 0.107 | 0.164 | 0.115 | 0.082 | 0.104 | - | 0.026 | 0.033 | 0.026 | 0.100 |
| MVe | 0.135 | 0.128 | 0.175 | 0.120 | 0.108 | 0.164 | 0.028 | - | 0.012 | 0.025 | 0.120 |
| Mho | 0.079 | 0.080 | 0.133 | 0.088 | 0.085 | 0.119 | 0.004 | 0.010 | - | 0.021 | 0.093 |
| MPr | 0.088 | 0.118 | 0.194 | 0.116 | 0.098 | 0.102 | 0.017 | 0.016 | 0.009 | - | 0.095 |
| NaA | 0.271 | 0.105 | 0.217 | 0.059 | 0.368 | 0.284 | 0.275 | 0.309 | 0.237 | 0.262 | - |

*Table 4.6   Genetic distances between 11 sheep breeds and populations.*

| | SAM | Dam | Dor | VaR | IFr | RoR | Mme | MVe | Mho | MPr |
|---|---|---|---|---|---|---|---|---|---|---|
| Dam | 0.553 | | | | | | | | | |
| Dor | 0.675 | 0.290 | | | | | | | | |
| VaR | 0.694 | 0.296 | 0.224 | | | | | | | |
| IFr | 0.273 | 0.604 | 0.496 | 0.582 | | | | | | |
| RoR | 0.532 | 0.422 | 0.514 | 0.470 | 0.700 | | | | | |
| MMe | 0.355 | 0.142 | 0.240 | 0.280 | 0.248 | 0.342 | | | | |
| MVe | 0.436 | 0.162 | 0.223 | 0.285 | 0.367 | 0.385 | 0.051 | | | |
| MHo | 0.434 | 0.191 | 0.163 | 0.242 | 0.319 | 0.437 | 0.079 | 0.022 | | |
| MPr | 0.276 | 0.181 | 0.233 | 0.312 | 0.272 | 0.375 | 0.062 | 0.047 | 0.046 | |
| NaA | 0.620 | 0.390 | 0.258 | 0.337 | 0.561 | 0.521 | 0.413 | 0.395 | 0.330 | 0.345 |

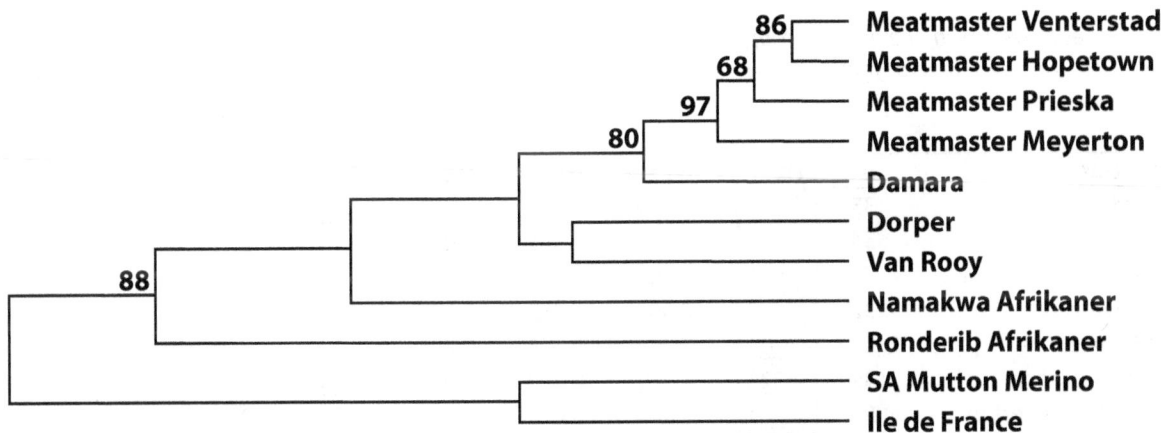

*Figure 4.5   Dendrogram based on genetic distances among 11 breeds and populations of sheep. Numbers at nodes indicate bootstrap support from 1000 replicates. Bootstrap values below 50% are not shown.*

A large proportion of individuals in each of these breeds (78.7-92.9%) were assigned to breed-specific clusters (Table 4.3). The majority (69.4%) of Meatmaster assignments were spread over three clusters, distinct from the clusters formed by established breeds, and with none of these three clusters containing individuals from one breeder only. While this assignment to three clusters (cluster 4, 6 and 9) may be interpreted as a partial lack of genetic identity among Meatmaster sheep, it should be noted that, while the three clusters contained 69.4% of Meatmaster sheep genotyped, none of these clusters contained more than 3.6% of individuals from any specific parent breeds. This implies that the Meatmaster indeed has genetic identity.

Trends from $R_{ST}$ values confirmed known relationships between breeds particularly well. Among-population $R_{ST}$ values among the four Meatmaster subpopulations were low, with higher values for pair-wise combinations of Meatmaster populations and the parent breeds, and still higher values among Meatmaster populations and the outgroup as expected (Table 4.5). Trends based on $F_{ST}$ were generally similar, although divergence between the Meatmaster populations and the outgroup were closer to values between the Meatmaster and parent breeds, compared to values from $R_{ST}$. These values thus provide strong support for the Meatmaster as a genetically recognizable group, with among population divergence being much lower than divergence between the Meatmaster populations and the parent breeds. The $F_{ST}$ values of 0.012-0.033 (Table 4.5) among Meatmaster populations are low compared to

representative published levels of differentiation between regional populations of sheep (for example 0.07 reported by Rendo et al., 2004 and 0.1 by Quiroz et al., 2008). The $R_{ST}$ values of 0.079-0.194 among Meatmaster populations and parent breeds are more closely comparable to the values published by these authors. Furthermore, genetic distance values among Meatmaster populations were much smaller than distance values between Meatmaster and parental breeds (Table 4.6), and the dendrogram constructed from distance values shows strong bootstrap support (97%) for a unique Meatmaster cluster (Figure 4.5). All these trends indicate uniqueness in the Meatmaster breed compared to its individual parent breeds.

Results from the hierarchical analysis of total genetic diversity provided additional support for the Meatmaster as a distinct group, with 2.315% of total variation found among Meatmaster populations compared to 11.445% found among breeds (including Meatmaster and all parent breeds). All the frequency-based measures used thus support the hypothesis that the Meatmaster is a genetically recognizable group.

The possible mechanism that may result in the formation of a genetically identifiable group or breed, following four independent selection programmes for the Meatmaster phenotype at four different localities, should be considered. The use of the Damara as the predominant parent breed by independent breeders most likely contributed to some identity between the end products of these separate selection programmes.

Table 4.7   Genetic diversity in one pooled group and 11 sheep breeds and populations, with sample size (n), unbiased heterozygosity (Hz), observed heterozygosity (Ho), and average number of alleles per locus (A). The abbreviation s.d. denotes standard deviation

| Population: | n | Hz | Hz s.d. | Ho | Ho s.d. | A | A s.d. |
|---|---|---|---|---|---|---|---|
| SA Mutton Merino | 35 | 0.704 | 0.030. | 0.631 | 0.026 | 6.70 | 2.41 |
| Damara | 34 | 0.665 | 0.040 | 0.585 | 0.028 | 6.00 | 2.11 |
| Dorper | 34 | 0.558 | 0.080 | 0.362 | 0.027 | 4.90 | 2.23 |
| Van Rooy | 32 | 0.599 | 0.067 | 0.475 | 0.029 | 4.80 | 1.99 |
| Ile de France | 40 | 0.705 | 0.037 | 0.496 | 0.025 | 6.90 | 2.23 |
| Ronderib Afrikaner | 35 | 0.524 | 0.074 | 0.499 | 0.027 | 4.30 | 1.57 |
| Meatmaster Meyerton | 55 | 0.705 | 0.047 | 0.616 | 0.021 | 7.70 | 2.21 |
| Meatmaster Venterstad | 40 | 0.634 | 0.057 | 0.541 | 0.025 | 6.70 | 2.00 |
| Meatmaster Hopetown | 40 | 0.675 | 0.051 | 0.519 | 0.025 | 8.20 | 2.10 |
| Meatmaster Prieska | 40 | 0.679 | 0.047 | 0.586 | 0.025 | 7.70 | 2.87 |
| Namaqua Afrikaner | 34 | 0.592 | 0.050 | 0.397 | 0.028 | 4.90 | 2.33 |
| Meatmaster Pooled | 175 | 0.689 | 0.049 | 0.570 | 0.012 | 9.9 | 2.38 |

## 4.5.1 Genetic diversity in the Meatmaster breed

Artificial selection may result in inbreeding depression and a reduction in fitness because relatively small population sizes are involved. The coefficients of genetic diversity used show that genetic diversity in individual Meatmaster populations is comparable to the higher end of the range of diversity values in the established parent breeds. Genetic diversity for the pooled population yielded values of Hz=0.689 for this breed compared to values of Hz=0.524-0.705 in other breeds (Table 4.7). The average heterozygosity in Meatmaster populations also compares favourably with values of 0.697, 0.620-0.728 and 0.738-0.769 calculated in sheep breeds by Arora and Bhatia (2004), Quiroz et al. (2008) and Rendo et al. (2004).

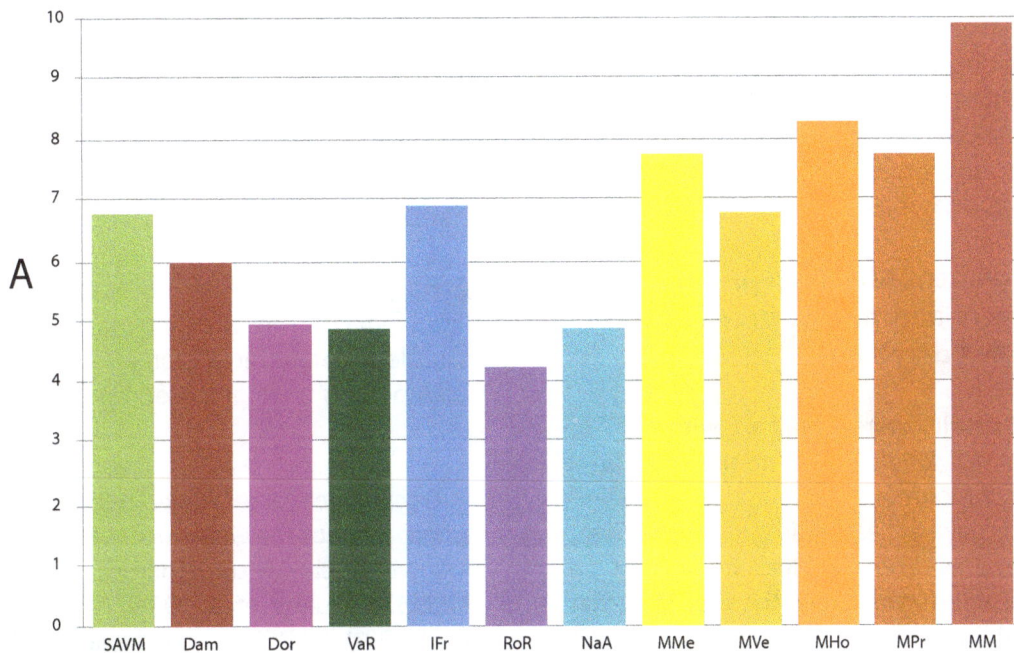

Figure 4.6   *Number of Alleles recorded for 12 sheep breeds/populations*

The A value in the Meatmaster was 9.9 (Figure 4.6), which is higher than the A=4.3-6.9 calculated for the parent breeds and also marginally higher than the upper limit of a typical range of values of 5.6-8.4 reported by Quiroz et al. (2008) in Mexican breeds. The high A value is particularly noteworthy since Spencer et al. (2000) has demonstrated that A may be the most accurate indicator of population bottlenecks. The high A value in the Meatmaster breed thus shows good retention of genetic diversity during the potential bottleneck created by selection in its early history. A large A also makes markers useful for future genetic screening in a breed (Kotze et al., 2004), for example during parentage studies.

The trend of high diversity in Meatmaster populations can be attributed to the origin and short history of the breed (10-15 years). The Meatmaster represents a mix of genetic material from five other breeds and a relatively high level of genetic diversity resulting from this admixture would be expected. Also note that the Damara breed, the biggest genetic contributor to the Meatmaster breed, has among the highest levels of diversity among the parent breeds. Furthermore, the Meatmaster is a young breed compared to the parent breeds and the Namaqua Afrikaner, and the time span of reproductive isolation and line-breeding within the breed may have been too short to have caused an appreciable loss of diversity. Management plans aimed at the conservation of genetic diversity should be implemented as the breed becomes established and full reproductive isolation from other breeds is established. Such measures should aim to find a balance between short-term gain and long-term sustainability.

## 4.6 Conclusions on Meatmaster DNA analysis

The Meatmaster breed shows promising levels of genetic identity for a young breed. A measure of definable uniqueness will be useful if it can eventually allow for molecular characterization of the breed. It will provide molecular corroboration for the official classification of the Meatmaster as a unique developing breed and it may have forensic applications (i.e. established Meatmaster breeders are experiencing the problem that non selected F1 crosses from parent breeds are bred and sold as *"Meatmaster"* sheep due to the fast growing popularity of the breed). Finally, results suggest that a high level of genetic diversity has been captured in the breed. The presence of high allelic diversity will reduce the likelihood of inbreeding depression in the short term and will be of benefit in providing diversity to use for specific applications such as parentage studies. The results from this project also contributed to the global genetic database.

## 4.7 Assignment of individuals applied to Meatmasters and parent breeds utilising the ARC DNA database

Significant results were obtained with the *"assignment of individuals"* based on the criterion of Rannala and Mountain (1997) and a 0.05 threshold, which gave a clear indication of the Meatmaster breed as distinctly different from its parent breeds and further more the four Meatmaster populations were either self classified (65.7%) or classified as other Meatmasters (29.5%), indicating common genetic factors between the Meatmaster populations and a clear difference between the Meatmaster breed and its parent breeds.

In total (132 samples) Meatmasters that classified as "Meatmasters" (Self or Other Meatmasters) were 95.3% and Meatmasters that classified as *"Parent Breeds"* (Damara and Van Rooy) were 4.7%. The SA Mutton Merino and Ronderib Afrikaner databases were 100.0% self classified while the Damara database had 8.6% individuals that classified as "Meatmasters", the Dorpers 8.8%, the Ile de France 5.0% and Van Rooy 3.1%.

## 4.8 Application of the Meatmaster DNA database at the ARC in the classification of the Gellapper sheep population in relation to the Meatmaster sheep breed

The first opportunity for the application of the Meatmaster DNA database presented itself early during 2009 when a request was received by FW Peters from the Department of Agriculture of Namibia to compare the DNA data from their Gellapper sheep, bred during research trials at Gelap-Ost research station near Keetmanshoop, to the Meatmaster DNA data and to advise them on the way forward.

### 4.8.1 Procedures followed to analyse Gellapper DNA

Genotypes of 10 unrelated males and 30 unrelated females from the Gellapper flock (Gel) were sampled. Whole blood samples were taken and DNA profiles were prepared by the ARC Genetics Lab at Irene.

Microsatellite markers were used for genetic characterization of the Gellapper sheep sampled. The following loci were used for the sheep genotyped: MCM527 (Hulme et.al, 1994), MAF65 (Buchanan et.al, 1991), OARFCB20 (Buchanan and Crawford, 1992), CSRD247 (Kemp et al., 1993), ETH225 (Steffen and Eggen, 1993), INRA63 (Vaiman et al., 1994) and TGLA53 (Crawford et al., 1995). Permission was obtained to make use of the existing DNA database of reference populations at the Agricultural Research Council (ARC, Irene) for the parent and other breeds. Preliminary statistical analysis of the Gellapper DNA data were performed at the University of the Free State in collaboration with Professor JP Grobler. To quantify levels of diversity within individual breeds and populations, observed heterozygosity (Ho), Nei's unbiased heterozygosity (Hz; Nei, 1987) and average number of alleles per locus (A) were calculated.

## 4.8.2  Results of Gellapper DNA analysis

*Table 4.8   F$_{ST}$ values for 8 sheep breeds compared to the Gellapper population*

**F$_{ST}$ values with Meatmasters pooled:**

|     | SAM | Dam | Dor | VaR | IFr | RoR | MM | NaA | Gel |
|-----|-----|-----|-----|-----|-----|-----|-----|-----|-----|
| SAM | 0 | | | | | | | | |
| Dam | 0.19131 | 0 | | | | | | | |
| Dor | 0.23770 | 0.10510 | 0 | | | | | | |
| VaR | 0.17889 | 0.06558 | 0.08629 | 0 | | | | | |
| IFr | 0.12009 | 0.17843 | 0.16604 | 0.1351 | 0 | | | | |
| RoR | 0.25419 | 0.13215 | 0.18064 | 0.1464 | 0.25730 | 0 | | | |
| MM | 0.15436 | 0.02764 | 0.04287 | 0.0368 | 0.13888 | 0.0955 | 0 | | |
| NaA | 0.18982 | 0.11999 | 0.20144 | 0.1329 | 0.15510 | 0.1856 | 0.1147 | 0 | |
| Gel | 0.32484 | 0.39072 | 0.44521 | 0.3893 | 0.34207 | 0.5226 | 0.3304 | 0.3902 | 0 |

## 4.8.3  Discussion of Gellapper DNA statistics

It is important for the future recognition and management of the developing Meatmaster breed to establish whether genetic differences between populations of this breed, and parent breeds, indicate real genetic distinctiveness or simply reflect random genetic drift (Peters et al., 2010).

When relatively small population sizes are involved artificial selection may result in inbreeding depression and a reduction in fitness. The coefficients of genetic diversity used show that genetic diversity in individual Meatmaster populations is comparable to the higher end of the range of diversity values in the established parent breeds. However genetic diversity for the Gellapper population yielded values of Hz=0.4290 for this population compared to values of Hz=0.5816-0.6960 in other Meatmaster populations and Hz=0.5413-0.6538 in the parent breeds. The average heterozygosity in the Gellapper population of 0.4290 was very low compared to values of 0.620-0.728 calculated in sheep breeds by Quiroz et al. (2008). This indicated a high level of inbreeding in the Gellapper population.

The F$_{ST}$ value of 0.3304 for the Gellapper population (Table 4.8) compared to the pooled value for Meatmasters indicated that the Gellapper were nearer to the Meatmaster breed than to its parent breeds for which F$_{ST}$ values of 0.3907-0.4452 were registered.

The A value (No of alleles) in the Gellapper population was 4.2 (Table 4.12), which is much lower than

the A=5.8-7.2 calculated for the parent breeds and also considerably lower than the A=6.8-8.6 for Meatmaster populations. The low A value is particularly noteworthy since Spencer et al. (2000) has demonstrated that A may be the most accurate indicator of population bottlenecks. The high A value in the Meatmaster breed shows good retention of genetic diversity (Peters et al., 2010) while the low A for the Gellapper indicates a possible population bottleneck.

## 4.8.4  Conclusion and recommendation in respect of Gellapper DNA analysis

Genetic diversity in the Gellapper sheep population was very low compared to diversity values in the parent breeds and other Meatmaster populations. F$_{ST}$ values however indicated that the Gellappers were closer to the Meatmaster breed than to any of the parent breeds. The low A value recorded for the Gellapper population also indicated a considerable amount of inbreeding. It was therefore proposed by FW Peters that the Gellapper be incorporated as a distinct Meatmaster population as part of the new Meatmaster sheep breed and that Meatmaster genetics from the South African Meatmaster populations be introduced, whereby the genetic variation of the Gellapper flock could be increased and the Meatmaster breed further expanded. The proposal was accepted by the Department of Agriculture in Namibia as well as the by the Meatmaster Sheep Breeders' Society of South Africa. Meatmaster rams bought at the Loeriesfontein national sale on 23 October 2009 were consequently introduced into the Gellapper flock at Gellap-Ost research station in Namibia.

Fig 4.7   (Left) The unique prolificant Gellapper line of Meatmaster sheep at Keetmanshoop. (Right) Freddie Peters, Rebekka Namwandi (Gellap-Ost), Clynton Collett and Brian Thawana  (Dept Agriculture Namibia) at the Loeriesfontein National sale (October 2009)

# The farmer's partner to knowledge resources

Visit www.kejafa.co.za
Email: kejafa@mweb.co.za
Tel: 011 025 4388

KefAfa
KNOWLEDGE WORKS

# Chapter 5   Characterization of the production potential of the Meatmaster sheep

## 5.1   Introduction

Schoeman et al., (2010) commented that: *"Approximately 80% of the agricultural land in South Africa is unsuitable for crop production, while the largest part thereof is not even suitable for either dairy or beef production. The small stock industry is consequently of crucial importance in the South African livestock environment."*

Much emphasis has been placed on the genetic characterization of livestock. Such genetic characterizations give important information on the genetic history of populations or the development of new or composite breeds, but do not render any information on the future utilization of the breed (Scholtz, 2005).

Campbell (1983) states that the following principles should be considered the most important in meat sheep production: Optimum health and weight gain in diverse agro-eco systems, optimum fertility and fecundity as well as acceptability of the product by its consumers. According to animal scientists breeding progress is seldom limited by the genetic composition of a breed but rather by the following factors:

- Breeders still tend to give too much attention to selection for traits such as colour and horns which bears no relation to productivity and furthermore attempt to select for too many traits simultaneously which results in very little breeding progress.

- As is evident from the minutes of many breed societies, there is a lack of consensus amongst breeders regarding their breeding objectives.

- Excessive value is attached to traits such as type and conformation which is not easily measurable and of which the heritability is generally low.

Breeding programmes designed to improve production efficiency require knowledge of genetic parameters for characters of economic importance such as growth rate, total weight of lamb weaned, prolificacy and reproduction (Matika et

al., 2001). Cloete et al. (2000) published a report on the productive performance of Dorper sheep and evaluated the phenotypes for productive traits such as fertility, fecundity and slaughter percentages. Snyman and Olivier (2002) compared the reproductive performance, growth, slaughter traits and a number of conformation characteristics of hair and wool type Dorper sheep under extensive conditions in the north-western Karoo region. The study concluded that there were no differences between hair and wool type Dorpers with respect to the economically important reproduction and growth traits.

*"As a measure of productive capacity, producing ability is of particular importance to commercial producers. Producing ability is neither a purely genetic value nor a purely environmental one it is a combination of both. Producing ability is a function of all those factors that permanently affect an individual's performance potential. Producing ability (PA) is therefore*

$$PA = BV + GCV + Ep$$

*culminating in the following genetic model for a trait in a performance record:*

$$P = \mu + BV + GCV + Ep + Et$$

*Where P denotes an individual's phenotype, $\mu$ the population mean phenotypic value for the trait, BV breeding value, GCV gene combination value, Ep permanent environmental effect and Et temporary environmental effect."* (Bourdon, 2000)

Adaptation of livestock to their production environments is important under natural production environments, particularly in the adverse South African environment. Unfortunately adaptation is complex and thus difficult to measure (Scholtz et al., 2010). However, several proxy-indicators for adaptation are available and have also been used (McManus et al., 2008). These include reproductive traits such as fertility, survival, birth rate and peri-natal mortality; production traits such as growth rate, milk production, low mortality and longevity; and health traits such as faecal egg counts and number of external parasites (Bonsma, 1980, 1983; Spickett et al., 1989; Scholtz et al., 1991; McManus et al., 2008).

**55**

*"Weaning weight is a possible selection criterion for early growth rate, however it has low direct heritability, maternal heritability and dam permanent environmental effect (0.19, 0.10 and 0.07, were respectively determined in the South African Dorper breed; Olivier and Cloete, 2006). Live weights at older ages have higher direct and maternal heritabilities (0.24 and 0.14 for South African Dorpers, respectively) it should therefore support sustained genetic progress in growth; however, they often have disadvantages as selection criteria because of preliminary culling and resultant small contemporary groups. It is therefore recommended to record an additional weight at approximately 270 or 365 days of age as a measurement of post-weaning growth."* (Schoeman et al., 2010). Snyman et al. (1993) characterized the productive and reproductive performance of Namaqua Afrikaner sheep and Snyman and Herselman (2005) compared the productive and reproductive performances of Afrino, Merino and Dorper sheep at different localities. No published information on the productive and reproductive performance of Meatmaster sheep is available.

*"How the measure of size and shape relate to the functioning of the individual is of paramount importance in livestock production. Therefore, constant checks on the relationships between body measurements and performance traits are vital in selection programmes."* (Maiwashe, 2000) The objective of the phenotypic data recorded is to provide an initial reference point for Meatmaster sheep from which future research can be done to determine relationships between performance and body measurements.

Substantial crossbreeding outlays involving multiple breeds were described by Erasmus et al. (1983) in the Southern Cape and by Greef et al. (1990) in Gauteng. Breed analysis of national flocks of sheep were reported for the Merino (Olivier and Cloete, 2007), Dohne Merino (Van Wyk et al., 2008) and the Dorper (Olivier and Cloete, 2006). The Meatmaster sheep breed was recently established (Government Gazette, 2007) as a locally developed sheep breed in South Africa. The primary objective of this chapter is the characterization of the new composite Meatmaster sheep in order for it to be established as a formal breed. There is still a shortage of available performance data at present but currently available data can provide an indication of the expected production potential of Meatmaster sheep.

The chapter consists of different datasets that are independent of each other. The results of each dataset will be presented separately to simplify the understanding and interpretation of the results.

## 5.2 Results and Discussion of Meatmaster production potential

Data recorded on the farm Schoongezicht on the Highveld in the Gauteng province, on the Gellap-Ost Research station at Keetmanshoop in Namibia, and on the farm La Rochelle in the Eastern Cape were used in the preparation of this chapter.

### 5.2.1 The Schoongezicht Highveld Meatmaster flock and parent breeds

Reproduction data recorded on the farm Schoongezicht near Meyerton on the Highveld in Gauteng between 1990 and 2006 comprising of 60 lifetime (7 years) reproduction records per breed for 3 sheep breeds in an extensive production environment. Data were collected to compare the reproduction potential of Meatmaster sheep with the reproduction potential of its parent breeds, in this instance Damara (paternal breed) and Ile de France sheep (maternal breed).

The phenotypic values for the 3 breeds were compared to each other for reproduction traits such as age at first lambing (A1stL), number of times lambed (NTL), number of lambs born per ewe that lambed (NLB), and inter lambing period (ILP). For all the breeds rams and ewes were kept in the flock in a free mating system. No supplementary feeding was supplied except rock salt. Lambs were not weaned by separation from the dams. Linear body measurements were also recorded on this flock and are discussed in this section.

The statistical procedure applied for analysis of the data sets in tables 5.1 and 5.2 was one way ANOVA (analysis of variance) in sequence of testing frequencies, anova and post hoc tests (Pallant, 2007). The Brown-Forsythe tests for equality of means were used to confirm the analysis of variance. The post-hoc tests applied differences of means to identify significant differences.

*Table 5.1   Comparison of reproductive traits between 3 breeds on the Schoongezicht farm in Gauteng*

| Trait | Breed | Number of records (N) | Mean | Standard Error (s.e.) | Minimum | Maximum |
|---|---|---|---|---|---|---|
| Age at first lambing (months) | Damara | 60 | 16.75[a] | 0.425 | 12 | 21 |
| | Ile de France | 60 | 23.20[a] | 0.638 | 16 | 37 |
| | Meatmaster | 60 | 14.6[a] | 0.306 | 11 | 22 |
| ILP (Inter-lambing period - days) | Damara | 60 | 266.32[a] | 9.52 | 180 | 454 |
| | Ile de France | 60 | 364.02[ab] | 5.77 | 265 | 441 |
| | Meatmaster | 60 | 273.03[b] | 9.13 | 177 | 426 |
| NTL (Lifetime number of lambings) | Damara | 60 | 8.00[a] | 0.26 | 4 | 11 |
| | Ile de France | 60 | 5.85[ab] | 0.12 | 4 | 8 |
| | Meatmaster | 60 | 7.80[b] | 0.29 | 4 | 12 |
| NLB (Lifetime number of lambs born) | Damara | 60 | 8.75[a] | 0.20 | 6 | 12 |
| | Ile de France | 60 | 8.00[a] | 0.18 | 5 | 11 |
| | Meatmaster | 60 | 9.57[a] | 0.24 | 6 | 15 |

[ab] *The mean difference is significant at the p<0.05 level for breeds with the same superscript.*

Age at first lambing (Table 5.1) were significantly (p<0.05) different for the three breeds and varied between 14.6 months for Meatmasters and 23.2 months for Ile De France ewes in these extensive conditions. The lower age at first lambing is a very important production trait and the lower phenotypic value for the Meatmaster is a unique characteristic. The ILP and NTL for the Meatmaster and the Damara were not significantly different but NLB for Meatmasters was significantly higher than the NLB for the Damara and the Ile De France.

For Highveld Meatmasters ILP (Table 5.1) varied between 177 and 426 days with a mean value of 273 days. Snyman et al. (1993) reported an average lambing interval for Namaqua Afrikaner sheep, run in a free mating system, of 274 days in the north-western Karoo.

On the Highveld farm fecundity values (lambs born per ewe that lambed) of 123% were recorded for Meatmasters, 109% for Damara and 136% for Ile de France sheep (Calculated from NTL and NLB in Table 5.1).

The reproductive performance of various age groups of the Highveld Meatmaster ewes of 5 to 8 years old were compared in respect of lifetime reproduction results. The age group of 5 year old ewes only included ewes of 5 years old, the same measure applied to the other groups. The data were collected on the farm Schoongezicht between 1996 and 2004. The results and statistics are presented in Table 5.2. (8 year old group born in 1996, 7 year old group born 1997, 6 year old group born 1998 and 5 year old group born in 1999)

**57**

*Table 5.2    Interlambing period, total births and total number of lambs born for different Meatmaster age groups on the Highveld*

| Trait | Age group | N | Mean | Std. Deviation | Std. Error | Minimum | Maximum |
|---|---|---|---|---|---|---|---|
| ILP (Inter-lambing period - days) | 5Y[a] | 60 | 235.50[c] | 59.74 | 7.713 | 170 | 409 |
| | 6Y[b] | 60 | 252.13 | 79.74 | 10.29 | 176 | 441 |
| | 7Y[c] | 60 | 273.03[a] | 70.69 | 9.13 | 177 | 426 |
| | 8Y[d] | 50 | 258.36 | 74.14 | 10.48 | 180 | 454 |
| NTL (Lifetime number of lambings) | 5Y[a] | 60 | 5.97 [bcd] | 1.45 | 0.19 | 3 | 8 |
| | 6Y[b] | 60 | 7.02 [ad] | 1.81 | 0.23 | 4 | 9 |
| | 7Y[c] | 60 | 7.80 [ad] | 2.22 | 0.29 | 4 | 12 |
| | 8Y[d] | 50 | 9.68 [abc] | 2.39 | 0.34 | 5 | 13 |
| NLB (Lifetime number of lambs born) | 5Y[a] | 60 | 7.25[cd] | 1.40 | 0.18 | 4 | 11 |
| | 6Y[b] | 60 | 7.75[cd] | 1.75 | 0.23 | 5 | 12 |
| | 7Y[c] | 60 | 9.57[abd] | 1.88 | 0.24 | 6 | 15 |
| | 8Y[d] | 50 | 11.36[abc] | 1.55 | 0.22 | 9 | 17 |

-[a] indicate p<0.05 compared to 5Y; -[b] indicate p<0.05 compared to 6Y;
 -[c] indicate p<0.05 compared to 7Y; -[d] indicate p<0.05 compared to 8Y

The NTL of the 6 year and 7 year old groups did not differ significantly but there was a significant difference between the NLB for the two groups due to more twins produced by the 7 year old group. The 8 year old group maintained its productivity with NTL and NLB values significantly higher than the other groups. The 7 year old group registered a higher fecundity value 123% but longer ILP (273.03 days), compared to the 6 year olds with a fecundity of 111% and an ILP of 252 days.

### 5.2.1.1   Linear body measurements of Meatmaster sheep

Linear body measurements (LBM) can be used in assessing growth rate, feed utilization and carcass characteristics in farm animals (Brown et al., 1973). Linear body measurements are divided into skeletal and tissue measurements (Essien and Adescope, 2003). The height at withers (WH) is part of skeletal measurements while the heart girth (HG) is part of tissue measurements (Blackmore et al., 1958). Several authors have established that testicular size is a good indicator of ram fertility (Schoeman, 1987; Duguma et al., 2002).

Age, location and pregnancy status seemed to have been the highest contributing factors to the variation in linear body measurements of Zulu sheep. The highest live weights for mature Zulu sheep (three and four pairs of incisors) were 39.76 kg and 40.26 kg with heart girths of 79.95 cm and 81.28 cm and wither heights of 67.36 and 68.02 cm, respectively (Kunene et al., 2007).

Phenotypic measurements of 137 Meatmaster sheep were recorded from the Highveld Meatmaster flock on the Schoongezicht farm near Meyerton in Gauteng. Meatmaster ewes (n = 110) and rams (n = 27), with four pairs of incisors or more, were used. The following linear body measurements were recorded using the measuring tape method as described by Fourie et al. (2002):

i.    Body weight (kg) (BW) – was taken with a sheep scale with 150 kg capacity and 200 gm precision.

ii.   Wither height (cm) (WH) - the highest point measured as the vertical distance from the top of the shoulder to the ground level.

iii.  Heart girth (cm) (HG) - the circumference of the chest posterior to the forelegs at right angles to the body axis.

iv. Canon bone (metacarpus) circumference (cm) (CC) - measured in the middle of the foreleg between the knee and pastern.

v. Canon bone (metacarpus) length (cm) (CL) - measured from the proximal to the distal epiphysis.

vi. Scrotal circumference (cm) (SC) - measured at the widest point of the scrotum (SC)

vii. Tail length (cm) (TL) – from the base to the tip of the tail.

viii. Head length (cm) (HL) – from the muzzle tip to the top of the head between the ears.

ix. Head width (cm) (HW) – in front across the eyes.

### 5.2.1.2 Discussion of Meatmaster body measurements

Statistical analysis was done with the one sample t-test (Norušis, 2003) to determine if there were significant differences between a number of measurements reported elsewhere in literature and linear body measurements recorded for Meatmaster sheep. For the other measurements means were obtained by ordinary summary statistics and standard deviations were calculated. The linear body measurements for the Highveld Meatmaster ewes and rams are summarised in Tables 5.3 and 5.4 respectively.

Table 5.3.  Meatmaster ewe body measurements

| LBM | N | Mean | Standard deviation (Standard error) | Min | Max |
|---|---|---|---|---|---|
| BW | 110 | 52.2 | 4.59 (0.44) | 46.0 | 64.0 |
| WH | 110 | 69.6 | 4.46 (0.43) | 64.5 | 81.5 |
| HG | 110 | 94.2 | 4.14 (0.40) | 87.5 | 101.5 |
| CC | 110 | 9.2 | 0.34 | 8.5 | 9.5 |
| CL | 110 | 14.4 | 1.39 | 12.0 | 16.0 |
| TL | 110 | 38.4 | 4.16 | 32.0 | 44.5 |
| HL | 110 | 23.0 | 1.12 | 21.5 | 24.5 |
| HW | 110 | 13.8 | 0.20 | 13.5 | 14.0 |

Table 5.4.  Meatmaster rams body measurements

| LBM | N | Mean | Standard deviation (Standard error) | Min | Max |
|---|---|---|---|---|---|
| BW | 27 | 58.1 | 4.94 (0.951) | 50.0 | 66.0 |
| WH | 27 | 74.6 | 3.81 (0.733) | 68.0 | 79.5 |
| HG | 27 | 92.6 | 3.96 (0.763) | 85.5 | 99.5 |
| CC | 27 | 10.0 | 0.37 (0.071) | 9.5 | 10.5 |
| CL | 27 | 15.7 | 0.94 (0.182) | 13.5 | 17.0 |
| SC | 27 | 33.4 | 2.16 (0.415) | 29.5 | 36.0 |
| TL | 27 | 40.6 | 5.75 | 32.5 | 49.5 |
| HL | 27 | 24.4 | 1.01 | 22.5 | 25.5 |
| HW | 27 | 14.6 | 0.53 | 14.0 | 15.5 |

Colour patterns, coat covering and hoof colors were noted. A wide range of colors and colour patterns were observed in Meatmaster sheep. Multi-coloured, uni-coloured or pied white, black, brown, tan, black and white patched, black and brown, dark-brown and grey. Colour patterns similarly show a wide spectrum some of which can be identified as white spotting, piebald spotting, white tail tip, persian pattern, turkish pattern, belted wading, flowery, raindrops, roan and combination patterns. Dark hoofs were observed in all the Meatmaster sheep.

Compared to Zulu (Nguni) sheep Meatmaster sheep are heavier, with greater wither height and heart girth values. For Zulu sheep of the same age, Kunene et al. (2007) determined BW = 34.72 kg, HG = 77.22 cm and WH = 63.66 cm. For Meatmaster ewes BW = 52.2 kg, HG = 94.2 cm and WH = 69.6 cm. One-sample T tests confirmed significant differences. For Meatmaster rams BW = 58.1 kg, HG = 92.6 cm, and WH = 74.6 cm. Since Zulu sheep is also a hair sheep type Meatmaster rams could be used on Zulu ewes to breed heavier lambs. Fourie et al. (2002) determined a number of phenotypic values for traits of young Dorper rams under extensive conditions recording 177 individuals from the Free State and 256 from the Northern Cape (BW = 57.5 kg and 50.8 kg respectively, HG = 92.8 cm and 88.3 cm, Canon bone circumferences (CC) = 10.3cm and 10.2cm, and Scrotal circumference (SC) = 33.0cm and 32.1cm). For Meatmaster rams no significant differences (p>0.05) were found between the traits of BW, HG, CC and SC compared to Dorper rams, but Meatmaster rams are significantly (p<0.05) taller than Dorper rams with WH = 63.7 cm and 62.9 cm and CL = 13.0 cm and 12.3 cm as measured by Fourie et al. (2002), which is indicated by the higher values for wither height (74.6 cm) and Canon bone length (15.7 cm) of the Meatmaster.

## 5.2.2 Eastern Cape Meatmaster flock and 4 other breeds at La Rochelle

Performance data were recorded in the Eastern Cape Province on the farm La Rochelle in the Venterstad district between 1990 and 2009 for 5 different sheep breeds in extensive conditions with no supplementary feeding. An average of 300 records per breed was used to compare phenotypic values for production traits such as age at first lambing, inter lambing period, fecundity, lamb mortality and weaning weight of lambs. This Meatmaster flock was developed from Damara and Dorper parent breeds with the Damara as the maternal breed and the Dorper as the paternal breed.

Performance records for Meatmaster sheep as recorded on the INTERGIS between 1999 and 2009 were statistically analysed with the SAS program. A total of 976 individual ewe lifetime reproduc-

tion records on age at first lambing, number of lambs weaned and number of lambs born, and 608 records on inter lambing period were used. The F generation (F1 to F5) of the Meatmaster sheep was also recorded. The procedures used are described in detail by SAS (1996).

The following model was applied for all traits:
$$Y_{ijkl} = \mu + t_i + g_j + k_k + (tg)_{ij} + b_1 BW + e_{ijkl}$$

Where
$Y_{ijkl}$ = trait of the l'th animal of the k'th number of lambing opportunities of the j'th F-generation of the ewe of the i'th birth year of the ewe,
$\mu$ = overall mean,
$t_i$ = fixed effect of the i'th year of birth of the ewe (1999 - 2008),
$g_j$ = fixed effect of the j'th F-generation of the ewe (F1 to F5),
$k_k$ = fixed effect of the k'th number of lambing opportunities (1 to 7),
$e_{ijkl}$ = random error with zero mean and variance $l\sigma^2_e$.

Descriptive statistics of reproductive data of Meatmaster ewes at La Rochelle are presented in Table 5.5, while the reproductive performance of ewes born from 1999 to 2008 are summarised in Table 5.6.

*Table 5.5 Descriptive statistics of reproductive data of Meatmaster sheep at La Rochelle*

| Statistic (976 records) | Number of lambing opportunities | Age at first lambing (days) | Inter lambing period (days) | Number of lambs born | Number of lambs weaned |
|---|---|---|---|---|---|
| Mean | 2.65 | 429.90 | 350.29 | 3.01 | 2.87 |
| Minimum | 1 | 259 | 177 | 1 | 0 |
| Maximum | 7 | 869 | 602 | 12 | 12 |
| Standard deviation | 1.75 | 85.17 | 39.53 | 2.26 | 2.26 |
| Standard error | 0.06 | 2.73 | 1.60 | 0.07 | 0.07 |
| Coefficient of variation | 66.17 | 19.81 | 11.28 | 75.05 | 78.66 |
| Skewness | 0.79 | 1.30 | 1.63 | 1.18 | 1.15 |
| Kurtosis | -0.70 | 2.00 | 12.10 | 0.74 | 0.72 |

Table 5.6  Reproductive performance (± s.e.) of ewes born from 1999 to 2008 in the flock at La Rochelle

| Year of birth | Number of lambing opportunities | Age at first lambing (days) | Inter lambing period (days) | Number of lambs born* | Number of lambs weaned* |
|---|---|---|---|---|---|
| 1999 | 3.97 ± 0.28 | 541 ± 15 | 429 ± 14 | 4.77 ± 0.14 | 4.53 ± 0.15 |
| 2000 | 3.52 ± 0.23 | 475 ± 12 | 400 ± 13 | 4.73 ± 0.12 | 4.52 ± 0.13 |
| 2001 | 3.28 ± 0.20 | 435 ± 11 | 363 ± 11 | 4.78 ± 0.11 | 4.66 ± 0.11 |
| 2002 | 2.52 ± 0.22 | 414 ± 12 | 335 ± 13 | 4.59 ± 0.12 | 4.39 ± 0.13 |
| 2003 | 3.97 ± 0.20 | 394 ± 11 | 342 ± 10 | 5.00 ± 0.10 | 4.80 ± 0.11 |
| 2004 | 3.27 ± 0.22 | 354 ± 12 | 345 ± 11 | 4.77 ± 0.11 | 4.62 ± 0.12 |
| 2005 | 2.80 ± 0.20 | 368 ± 12 | 342 ± 11 | 4.75 ± 0.11 | 4.55 ± 0.12 |
| 2006 | 2.37 ± 0.20 | 372 ± 12 | 319 ± 11 | 4.72 ± 0.11 | 4.18 ± 0.12 |
| 2007 | 1.63 ± 0.16 | 403 ± 10 | 328 ± 10 | 4.72 ± 0.10 | 4.64 ± 0.10 |
| 2008 | 0.96 ± 0.16 | 404 ± 11 | - | 4.81 ± 0.10 | 4.72 ± 0.11 |

* Number of lambs born and weaned over 2.65 lambing opportunities

Age at first lambing for Meatmaster sheep in the Eastern Cape (Table 5.6), varied from 354 to 541 days (11.6 – 17.8 months) with a mean value of 429.9 days (14.1 months). For ewes born from 1999 to 2004, A1stL gradually declined from 541 (17.8 months) to 354 (11.6 months) in 2004 and for 2005 to 2008 ewes, the age at first lambing varied between 367 and 404 days (12 – 13.3 months).

The lamb survival percentages varied between 95.0% and 98.3% for the ewes born from 1999 to 2008 (Table 5.7) except for the ewes born in 2006, which registered a low value of 88.6%. The 2003 ewes were the best performers for number of lambs weaned with the 2008 ewes the second best performers. The age at first lambing decreased constantly from 1999 ewes to 2004 ewes, 541 to 354 days (17.8 to 11.6 months) and there after gradually increased from 368 to 404 days (12.1 to 13.2 months). The shortest lambing interval recorded was 319 days (10.5 months) for the 2006 ewes which also had the worst lamb survival rate (88.6%), but in contradiction to the 2006 group the 2007 ewes which had the second lowest inter lambing period 328 days (10.8 months) registered the highest lamb survival rate (98.3%). Due to the fact that the ewes were not kept in a free mating management system the ILP does not bear relationship with the production potential of the Meatmaster ewes since this trait was artificially controlled. The high age at first lamb (A1stL) of the 1999 born ewes can likewise be explained. A1stL values of between 11.6 to 13.2 months for most of the groups, as well as lamb survival rates of 95.6% to 98.3% are very encouraging indicators of the production potential of the Meatmasters.

Fecundity (number of lambs born per 100 ewes that lambed), weaning percentage (number of lambs weaned per 100 ewes that lambed) and survival rate (number of lambs weaned per number of lambs born of ewes that had different numbers of lambing opportunities) are summarised in Table 5.7.

Table 5.7.  Reproductive performance of Meatmaster ewes that had different numbers of lambing opportunities (976 records)

| Number of lambing opportunities | Fecundity (%) | Weaning (%) | Survival rate (%) |
|---|---|---|---|
| 1 | 106.0 | 89.3 | 84.2 |
| 2 | 109.0 | 99.0 | 90.8 |
| 3 | 111.0 | 108.0 | 97.3 |
| 4 | 115.0 | 112.0 | 97.4 |
| 5 | 116.0 | 112.0 | 97.0 |
| 6 | 119.0 | 116.0 | 97.0 |
| 7 | 132.0 | 126.0 | 95.5 |
| Mean | 115.4 | 108.9 | 94.2 |

From results obtained in the Eastern Cape (Table 5.7) for ewes that lambed 1 to 7 times, a steady increase in fecundity from 106% to 132% was obtained with an average phenotypic value of 115.4%.

Snyman and Olivier (2002) reported lamb survival rates of 96.9% and 95.79% for hair and wool type Dorper sheep respectively. In a study of 3 breeds in extensive conditions in the Eastern Cape, Snyman and Herselman (2005) reported lamb survival rates of 77.0% to 82.1% for Dorpers, 87.1% to 91.2% for Afrinos and 85.1% to 90.5% for Merino sheep. Matika et al. (2003) reported lamb survival of 85% for Sabi lambs, while Tibbo (2006) reported a lamb survival rate of 74.6% for Horro lambs. The survival rate recorded for Meatmaster lambs compares favourably with the highest values reported in literature. Lamb survival percentages increased from 84.2% to 97.4% for ewes with one to ewes with four lambing opportunities.

The effect of generation number on reproductive performance of ewes born from 1999 to 2008 in the flock is presented in Table 5.8. A1stL for the F1 generation (Table 5.8) was 407.7 days and for the F5 generation 406.6 days.

*Table 5.8. Effect of generation number on reproductive performance of ewes born from 1999 to 2008 in the flock at La Rochelle*

| Generation | Age at first lambing (days) | Inter lambing period (days) | Number of lambs born* | Number of lambs weaned * |
|---|---|---|---|---|
| F1 | 408 ± 9 | 289 ± 11 | 4.71 ± 0.08 | 4.54 ± 0.09 |
| F2 | 428 ± 6 | 340 ± 8 | 4.80 ± 0.06 | 4.62 ± 0.07 |
| F3 | 432 ± 7 | 333 ± 9 | 4.78 ± 0.07 | 4.61 ± 0.08 |
| F4 | 407 ± 10 | 333 ± 12 | 4.75 ± 0.10 | 4.51 ± 0.10 |
| F5 | 407 ± 27 | 376 ± 32 | 4.79 ± 0.26 | 4.43 ± 0.26 |

*\* Number of lambs born and weaned over 2.65 lambing opportunities*

For most of the traits compared A1stL, NLB and NLW performance were fairly constant between the generations F1 to F5 (Table 5.8) which seem to indicate good retention of hybrid vigour. Van Wyk et al. (2009) concluded from a study on Dormer sheep where data were analysed to quantify the increase in actual level of inbreeding and to investigate the effect of inbreeding on phenotypic values, genetic parameters and estimated breeding values that no significant inbreeding depression occurred over a period of 62 years in the closed Elsenburg flock, except for -0.006 kg for birth weight and -0.093 kg for weaning weight. This finding by van Wyk et al. (2009) confirms the theory by Bourdon, (2000) that no hybrid vigour is lost after the F2 generation provided that the popula-

tion size is large enough or alternatively that the dominance model of hybrid vigour is appropriate. No conclusion can be made regarding ILP which was artificially controlled.

Data collected from 2000 tot 2009 on the farm La Rochelle were used for the analysis of weaning and post-weaning weights. Data were obtained from the INTERGIS. A total of 3406 records for weaning weight (ww) (corrected for age of the lamb) and 1120 records for 270 day weight (pw) (corrected for age) were used.

The following model was applied for weaning weight and post weaning weight:
$$Y_{ijkl} = \mu + t_i + g_j + k_k + (tg)_{ij} + b_1BW + e_{ijkl}$$
Where
$Y_{ijkl}$ = trait of the l'th animal of the k'th birth status of the j'th sex of the i'th birth year,
$\mu$ = overall mean,
$t_i$ = fixed effect of the i'th year of birth (2000 - 2009),
$g_j$ = fixed effect of the j'th sex (Male, Female),
$k_k$ = fixed effect of the k'th birth status (1, 2, 3),
$(tg)_{ij}$ = effect of the interaction between the j'th sex and the k'th birth status,
$e_{ijkl}$ = random error with zero mean and variance I$\sigma^2_e$.
The effect of sex and birth status of the lamb on weaning and post-weaning weight of Meatmaster lambs at La Rochelle are summarised in Table 5.9.

*Table 5.9 Weaning and post-weaning weight (± s.e.) of Meatmaster lambs at La Rochelle*

| | Weaning weight (kg) Mean = 24.92 kg | Post-weaning weight (kg) Mean = 46.86 kg |
|---|---|---|
| Ram lambs | 27.09[a] ± 0.38 | 46.16[a] ± 0.7 |
| Ewe lambs | 24.40[a] ± 0.44 | 47.44[a] ± 0.81 |
| Singles | 25.59[b] ± 0.10 | 48.01[b] ± 0.39 |
| Twins | 24.65[bc] ± 0.16 | 44.98[b] ± 0.46 |
| Triplets | 26.99[c] ± 0.85 | 47.42 ± 1.51 |
| Ewe single | 24.73 ± 0.13 | 47.98 ± 0.41 |
| Ewe twin | 23.39 ± 0.20 | 44.80 ± 0.52 |
| Ewe triplet | 25.07 ± 1.30 | 49.56 ± 2.19 |
| Ram single | 26.45 ± 0.12 | 48.03 ± 0.42 |
| Ram twin | 25.90 ± 0.22 | 45.16 ± 0.59 |
| Ram triplet | 28.91± 1.08 | 46.28 ± 2.02 |

*[a, b, c] Values within fixed effect with the same superscript differed significantly at the P<0.05 level*

100-day mean weaning weights recorded in the Eastern Cape varied from 21.74kg to 29.19kg between 2000 and 2009 indicating a strong environmental effect on the phenotypic value for the trait of weaning weight. From Table 5.9 mean weaning weights for ram lambs were 27.09 kg, ewe lambs 24.4 kg, singles 25.59 kg, twins 24.65 kg and triplets 26.99 kg. These values compare well to 100-day weaning weights reported by Snyman et al. (1993) for Namaqua Afrikaner sheep, which was reported as 26.09 kg for ram lambs and 24.74 kg for ewe lambs.

270-day mean weights varied between 43.5 kg and 51.43 kg during 2003 to 2009 with a mean value for males of 46.16 kg and for females 47.44 kg. Singles had a mean phenotypic value of 48.01 kg, twins 44.99 kg and triplets 47.42 kg (Table 5.9).

Snyman and Herselman (2005) reported 270-day weights for Dorper lambs at two locations in the Eastern Cape as 43.4 kg and 41.3 kg. Reporting on the comparison of hair and wool type Dorper sheep Snyman and Olivier (2002) reported 270-day weights of 48.2 kg and 47.8 kg respectively.

270-day weights for Meatmasters compare well to the higher end of values reported for Dorper sheep.

### 5.2.2.1 Meatmaster genetic trends: La Rochelle Eastern Cape (1999-2009)

Breeding values for weaning weight, post weaning weight, lifetime total weight of lamb weaned and relative economic value were obtained from the INTERGIS.

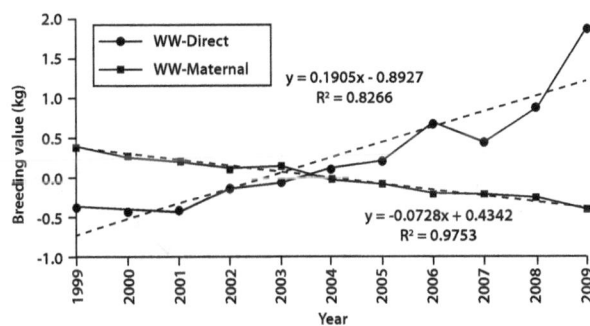

Figure 5.1  Direct and maternal genetic trends in weaning weight (WW)

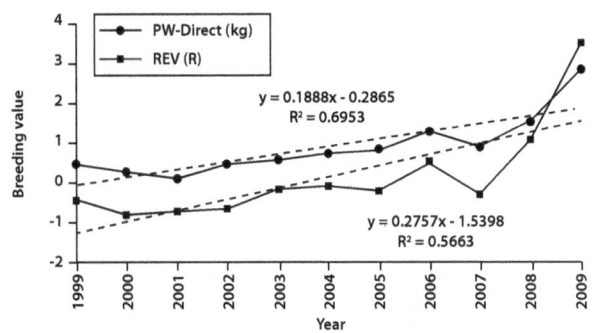

Figure 5.2 Genetic trends in post-weaning weight (PW) and relative economic value (REV)

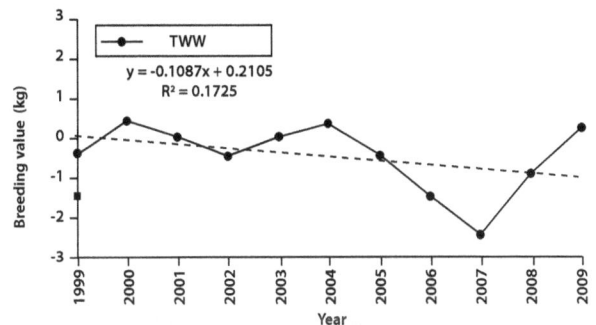

Figure 5.3  Genetic trends in total weight of lamb weaned (TWW)

Meatmaster genetic trends calculated from available Meatmaster performance data in the Eastern Cape (Fig 5.1) indicates a constant increase in direct WW breeding values with a limited decrease in maternal breeding values.

PW direct breeding values gradually increased (Fig 5.2) and most importantly the revenue (R) breeding values increased constantly.

TWW breeding values decreased slightly (Fig 5.3). The graph indicates a downward trend from 2004 to 2007 but a steep increase from 2007 to 2009. If the upward trend can be maintained further improvement can be expected.

### 5.2.2.2  Comparison of 5 breeds at La Rochelle

Data collected on the farm La Rochelle between 1990 and 2004 were used for this analysis. A total of 1517 records were used of which 302 were for SAMM, 315 for Afrino, 325 for Meatmaster, 295 for Dorper and 280 for Damara sheep. All these records were used for age at first lambing. For the traits of ILP, WW for singles and WW for twins the number of records used were as tabulated in Table 5.10.

The statistical procedure applied for analysis of the data in table 5.10 was one-way anova (analysis of variance) in sequence of testing frequencies, ANOVA and post hoc tests (Pallant, 2007). The

Brown-Forsythe tests for equality of means were used to confirm the analysis of variance. The post-hoc tests apply differences of means to identify significant differences.

*Table 5.10   Descriptive statistics of reproductive data and weaning weights of five breeds at La Rochelle, Eastern Cape (1990 – 2004)*

| Trait | Breed | Number of records (N) | Mean | Standard Error (s.e.) | Minimum | Maximum |
|---|---|---|---|---|---|---|
| A1stL | SAMM[a] | 302 | 22.00 [bcde] | 0.26 | 16 | 36 |
| | Afrino[b] | 315 | 18.01 [acde] | 0.13 | 16 | 24 |
| | MeatM[c] | 325 | 14.01 [abde] | 0.09 | 11 | 18 |
| | Dorper[d] | 295 | 19.01 [abce] | 0.23 | 12 | 26 |
| | Damara[e] | 280 | 15.01 [abcd] | 0.13 | 10 | 18 |
| ILP | SAMM[a] | 285 | 412.00 [bcde] | 1.24 | 365 | 459 |
| | Afrino[b] | 306 | 58.05 [ace] | 2.07 | 240 | 434 |
| | MeatM[c] | 307 | 349.04 [abde] | 1.49 | 180 | 365 |
| | Dorper[d] | 279 | 357.01 [ace] | 2.04 | 240 | 434 |
| | Damara[e] | 263 | 270.05 [abcd] | 2.65 | 180 | 365 |
| WWsl | SAMM[a] | 211 | 31.85[bde] | 0.37 | 19.80 | 44.00 |
| | Afrino[b] | 203 | 29.35 [ace] | 0.33 | 21.50 | 40.50 |
| | MeatM[c] | 223 | 26.95[abde] | 0.29 | 19.60 | 39.20 |
| | Dorper[d] | 175 | 30.31 [ce] | 0.35 | 19.80 | 42.50 |
| | Damara[e] | 170 | 22.15[abcd] | 0.26 | 14.10 | 34.50 |
| WWtl | SAMM[a] | 122 | 23.05 [e] | 0.26 | 17.00 | 33.50 |
| | Afrino[b] | 113 | 22.30[de] | 0.21 | 18.10 | 31.50 |
| | MeatM[c] | 98 | 22.80[de] | 0.18 | 17.60 | 33.00 |
| | Dorper[d] | 119 | 24.11[bce] | 0.25 | 18.50 | 34.50 |
| | Damara[e] | 112 | 17.25[abcd] | 0.28 | 9.20 | 29.60 |

-[a] indicate p<0.05 compared to SAMM; -[b] indicate p<0.05 compared to Afrino;
-[c]indicate p<0.05 compared to MeatM; -[d] indicate p<0.05 compared to Dorper;-[e] indicate p<0.05 compared to Damara

For all the breeds recorded at La Rochelle in the Eastern Cape (Table 5.10) a significant difference was recorded for the phenotypic values for the production trait of A1stL. Mean values for Meatmasters were 14.01 months; Damara 15.01 months; Afrino 18.01 months; Dorper 19.01 months and SA Mutton Merino 22.00 months. Age at first lambing is a very crucial factor in any commercial small stock enterprise. The lower the A1stL value the greater is the potential for genetic improvement of a population due to the shorter generation in-

terval. Generation interval is the amount of time required to replace one generation with the next. The shorter the generation interval, the faster the rate of genetic change (Bourdon, 2000).

Where weaning weights were compared on the same farm for 5 breeds (Table 5.10) Mean weaning weights for single Meatmaster lambs were 26.95 kg compared to 29.35 kg for Afrinos, 30.0 kg for Dorper lambs and 22.15 kg for Damara lambs. Meatmaster twins were recorded as 22.8 kg compared to 22.3 kg for Afrinos 24.1 kg for Dorper lambs and 17.25 kg for Damara lambs. Weaning weights for Afrino lambs were reported by Snyman and Herselman, (2005) at two localities as 28.1 kg and 26.9 kg respectively and Dorper lambs as 31.0 kg and 30.7 kg respectively.

Although the ILP value for Meatmasters of 349.04 days differed significantly from the ILP of 357.01 days for Dorpers and the ILP of 270.05 days for Damaras the control of lambing seasons institutes artificial control on these values and it cannot be used to characterize the natural potential for this specific trait. The shorter ILP value for the Damara is in this case attributable to the fact that the Damara breed was mated to lamb three times in two years while the other breeds were mated to lamb once a year.

### 5.2.3 Keetmanshoop Meatmaster (Gellapper) flock and parent breeds

Performance was recorded by the Gellap-Ost Research Station at Keetmanshoop in Namibia for three breeds, Meatmaster (Gellapper), Damara and Dorper between 2003 and 2004 for an average of 55 ewes per breed. Phenotypic values for productive traits such as lambs born per ewe mated, fecundity, lambs weaned, weaning weight, slaughter weight and carcass grading were recorded. All the flocks were run simultaneously in extensive conditions without supplementary feeding. Mating was done only once a year for a single lambing season per year.

*Table 5.11  Descriptive statistics of five Meatmaster (Ge) production traits at Keetmanshoop*

| Statistic | Birth weight (kg) | Weaning weight (kg) | ADG (g/d) | Live slaughter weight LSW (kg) | Cold carcass weight CCW (kg) |
|---|---|---|---|---|---|
| Mean | 3.35 | 24.89 | 149.54 | 35.95 | 15.64 |
| Minimum | 2.80 | 19.00 | 110.00 | 29.00 | 12.4 |
| Maximum | 4.40 | 38.00 | 206.00 | 46.50 | 20.5 |
| Standard deviation | 0.49 | 4.29 | 25.51 | 4.57 | 1.80 |
| Standard error | 0.06 | 0.53 | 3.14 | 0.56 | 0.22 |

*Table 5.12  Comparison of five production traits for 3 breeds at Keetmanshoop*

| Trait | Breed | Number of records (N) | Mean | Standard Error (s.e.) | Minimum | Maximum |
|---|---|---|---|---|---|---|
| Birth weight (kg) | MMGel | 66 | 3.35 [a] | 0.06 | 2.8 | 4.4 |
|  | Damara | 41 | 3.42 [b] | 0.03 | 2.9 | 3.8 |
|  | Dorper | 44 | 3.70 [ab] | 0.06 | 2.9 | 4.1 |
| Weaning weight (kg) | MMGel | 54 | 24.89 [a] | 0.53 | 19.00 | 38.00 |
|  | Damara | 34 | 18.62 [ab] | 0.28 | 16.00 | 25.00 |
|  | Dorper | 38 | 25.34 [b] | 0.56 | 18.70 | 32.00 |
| ADG (g/d) | MMGel | 54 | 149.54 [a] | 3.14 | 110.0 | 206.0 |
|  | Damara | 34 | 110.82 [ab] | 2.49 | 103.0 | 180.0 |
|  | Dorper | 38 | 149.41 [b] | 2.56 | 110.0 | 197.0 |
| Live slaughter weight (kg) | MMGel | 53 | 35.95 [a] | 0.56 | 29.00 | 46.50 |
|  | Damara | 32 | 29.67 [ab] | 0.37 | 25.00 | 36.00 |
|  | Dorper | 36 | 36.85 [b] | 0.54 | 29.00 | 45.00 |
| Carcass weight (kg) | MMGel | 53 | 15.64 [a] | 0.22 | 12.40 | 20.50 |
|  | Damara | 32 | 13.05 [ab] | 0.17 | 11.50 | 15.70 |
|  | Dorper | 36 | 15.73 [b] | 0.22 | 13.00 | 19.50 |

[a, b] *The mean difference is significant at the p<0.05 level for breeds with the same superscript.*

Birth weight recorded for Meatmaster lambs at Keetmanshoop in Namibia (Table 5.12) varied from 2.8 kg to 4.4 kg with a mean value of 3.35 kg compared to 3.42 kg for Damara lambs and 3.7 kg for Dorper lambs. A high percentage (53%) of twins was included in the Meatmaster group. Snyman and Olivier (2002) reported birth weights for hair and wool type Dorpers as 4.06 kg and 4.12 kg respectively. Matika et al. (2003) reported birth weights for Sabi lambs as 2.63 kg. Tibbo (2006) reported birth weights of 2.40 kg for Horro and 2.06 kg for Menz lambs. El-Arian et al. (2008) reported birth weights of 2.53 kg for Romanov lambs in Egypt. The values for Meatmaster lambs compare well to the higher end of values reported in literature.

At Keetmanshoop (Table 5.12) weaning weights for Meatmasters were recorded as 24.88 kg compared to 25.34 kg for Dorper lambs and 18.62 kg

for Damara lambs. Weaning weights of 12.8 kg were reported for Sabi sheep, 10.31 kg for Horro, 9.51 kg for Menz and 12.03 kg for Romanov lambs (Matika et al., 2003; Tibbo, 2006 and El-Arian et al., 2008). The Keetmanshoop values for Meatmaster lambs were well within the upper range of weaning weights reported in literature.

Percentage ewes lambing per ewes mated for Meatmaster (Ge) sheep at Keetmanshoop (Table 5.13) Meatmaster (Gellapper) sheep was recorded at 86% compared to 74.5% for Damara and 69.9% for Dorper sheep in the same trail. Snyman and Olivier (2002) reported lambing percentages (percentage ewes lambing per ewes mated) of 85.9% for hair type Dorper sheep and 82.5 % for wool type Dorpers. Lambing percentages of 85.95% was reported for Namaqua Afrikaner sheep (Snyman et al., 1993).

*Table 5.13   Lamb production recorded for 3 breeds at Keetmanshoop (2003-2004)*

|  | Meatmaster (Ge) | Damara | Dorper |
|---|---|---|---|
| Mating season | 18/04/2003 – 25/05/2003 | | |
| Number of ewes mated | 50 | 51 | 55 |
| Average weight of ewes at lambing (kg) | 55 [a] | 52 [a] | 57 [a] |

[a] *Estimated value. True values for the 2003 mating season were not obtained.*

*Table 5.13   Continued*

|  | Meatmaster (Ge) | Damara | Dorper |
|---|---|---|---|
| Lambing season | 01/9/03-07/10/03 | 22/9/03-20/10/03 | 17/9/03-17/10/03 |
| Number of ewes that lambed | 43 | 38 | 38 |
| Lambing % (% ewes lambed per ewes mated) | 86.00 | 74.51 | 69.09 |
| Fecundity (% lambs born per ewes that lambed) | 153.50 | 107.89 | 115.79 |
| Number of lambs born | 66 | 41 | 44 |
| Lamb% (% lambs born per ewes mated) | 132.00 | 80.39 | 80.00 |
| Average birth weight (kg) & No. of lambs weighed at birth | 3.35<br>63 | 3.42<br>41 | 3.70<br>44 |

*Table 5.14   Fecundity and weaning results (Keetmanshoop)*

|  | Meatmaster (Ge) | Damara | Dorper |
|---|---|---|---|
| Weaning date | 19/02/2004 | | |
| Average age at wean (days) | 143 | 138 | 144 |
| Number of lambs weaned | 54 | 34 | 38 |
| Weaning % (% lambs weaned per ewes mated) | 108.00 | 66.67 | 69.09 |
| Average weaning weight (kg) | 24.88 | 18.62 | 25.34 |
| Total weaned lamb weight (kg) | 1 343.75 | 633.00 | 962.75 |
| Average wean lamb weight / ewe mated (kg) (Economic turnover) | 24.88 | 12.41 | 17.50 |

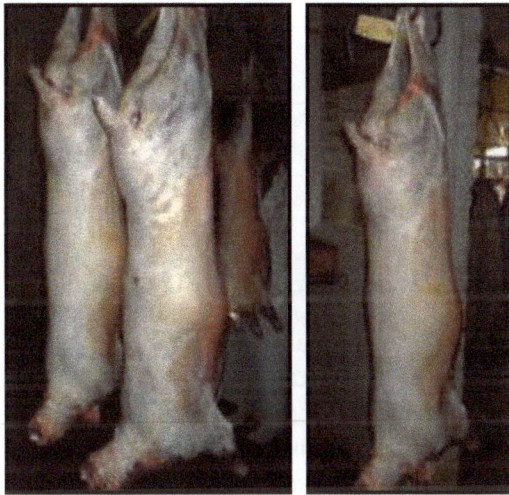

*Fig 5.4   Meatmaster carcasses achieve good gradings and are acceptable to the market*

*Table 5.15   Slaughter weight and age of trail lambs (Keetmanshoop)*

|  | Meatmaster (Ge) | Damara | Dorper |
|---|---|---|---|
| Slaughter date | 28/04/2004 | | |
| Number of lambs | 53 | 32 | 36 |
| Average age at slaughter for trial lambs (days) | 212 | 207 | 214 |
| Average live slaughter weight for trial lambs (kg) | 35.95 | 29.67 | 36.85 |

*Table 5.16 Carcass grade and weight of lambs (Keetmanshoop)*

|  | Meatmaster (Ge) | Damara | Dorper |
|---|---|---|---|
| Average warm carcass weight for trial lambs (kg) | 16.13 | 13.05 | 16.61 |
| Average cold carcass weight for trial lambs (kg) | 15.64 | 13.05 | 15.73 |
| Slaughter % (% Cold carcass weight / Live slaughter weight) | 43.51 | 42.66 | 43.71 |
| Average Build (Conformation) for trial lambs | 2.75 | 2.19 | 3.72 |
| Average fat count for trial lambs – All | 1.69 | 1.84 | 2.11 |
| % of trial lambs per grade -Grade A0 | 15.09 | 0 | 0 |
| Grade A1 | 20.75 | 31.25 | 5.56 |
| Grade A2 | 50.94 | 50.00 | 80.56 |
| Grade A3 | 9.43 | 18.75 | 11.11 |
| Grade A4 | 1.89 | 0 | 2.78 |
| Grade A5 | 1.89 | 0 | 0 |
| Average price per grade - Grade A0 (N$/kg) | 13.50-14.50 | - | - |
| Grade A1 (N$/kg) | 16.00 | 14.00-15.00 | 15.00-16.00 |
| Grade A2 (N$/kg) | 19.00-19.50 | 17.00-18.00 | 19.00-19.50 |
| Grade A3 (N$/kg) | 19.00-19.50 | 18.00 | 19.50 |
| Grade A4 (N$/kg) | 16.00 | - | 16.00 |
| Grade A5 (N$/kg) | 14.20 | - | - |
| Average income per lamb (N$) | 277.98 | 212.80 | 307.51 |

*Table 5.17   Gross economic returns (2004, N$)  (Keetmanshoop)*

|  | Meatmaster (Ge) | Damara | Dorper |
|---|---|---|---|
| Total lambs slaughtered per group | 53<br>(1 dead after wean) | 32<br>(1 dead after wean) | 36<br>(2 dead after wean) |
| Total cold carcass weight (kg) | 829.40 | 405.10 | 579.80 |
| Average cold carcass weight /<br>ewe mated (kg) | 16.59 | 7.94 | 10.54 |
| Average income per ewe mated (N$)<br>(for all slaughtered lambs) | 294.66 | 133.52 | 201.28 |
| Average weight of adult ewe (kg) | 55 | 52 | 57 |
| Total ewe weight (kg) | 2 750 | 2 652 | 3 135 |
| Number of ewes per hectare (ewes/ha)<br>(at 7.5kg biomass / ha stocking rate) | 0.136 | 0.144 | 0.132 |
| Amount of hectares needed for<br>number of ewes mated (ha) | 367.6 | 354.2 | 416.7 |
| Income / hectare (Total income/<br>Amount of hectares needed for group)<br>(N$/ha) | 40.08 | 19.23 | 26.57 |

The % lambs weaned per ewes mated (Table 5.14) of 108% for the Meatmaster (Ge) was consequently well above the 66.7% of the Damara and 69.1% of the Dorper resulting in 24.88kg lamb weaned per Meatmaster (Ge) ewe compared to 12.41kg for the Damara and 17.5kg for the Dorper. The Meatmaster (Ge) ewes finally produced 16.59 kg of carcass weight per ewe mated (Table 5.17) compared to the 7.94 kg/ewe mated for the Damara and 10.54 kg/ewe mated for the Dorper breed.

In the Keetmanshoop trial for the September/October 2003 lambing season (Table 5.13) a fecundity value of 153.5% was recorded for the (Gellapper) Meatmasters as well as 107.9% for Damara and 115.8% for the Dorper sheep. From September 2001 to March 2003 Snyman and Herselman (2005) recorded fecundity values of 129.9% and 139.7% for Dorper sheep at two different localities in the Eastern Cape.

From Tables 5.15 and 5.16 live slaughter weights (LSW) were recorded for Meatmaster, Damara and Dorper lambs as 35.95 kg, 29.67 kg and 36.85 kg respectively. Corresponding CCW values were 15.64 kg, 13.05 kg and 15.73 kg respectively resulting in slaughter percentages of 43.51%, 42.66% and 43.71%.

Snyman and Herselman (2005) reported live slaughter weights of 40.9 kg and 39.2 kg for Dor-per lambs at 248 days and 295 days in the Eastern Cape, with CCW values of 19.0 kg and 19.2 kg resulting in dressing percentages of 45.5% and 46.8% respectively. Matika et al. (2003) reported slaughter weights of 29 kg with CCW of 13.1 kg for Sabi sheep in Zimbabwe. Tibbo (2006) reported LSW for Horro and Menz sheep at 18 months as 25 kg with resulting CCW of 10.0 kg at a dressing percentage of 40%.

At the Keetmanshoop  trial carcass gradings for Meatmaster lambs (Table 5.16) were spread from A0 to A5 with 15.09% A0, 20.75% A1, 50.94% A2, 9.43% A3, 1.89% A4 and 1.89% A5 carcasses. Damara lambs graded 31.25% A1, 50.0% A2 and 18.75% A3, while Dorper lambs graded 5.56% A1, 80.56% A2, 11.11% A3 and 2.78% A4. The high percentage of twin lambs among the Meatmasters were slaughtered at the same age during the trials while they were not really in the required condition for slaughter. Snyman and Olivier (2002) reported a spread of carcass gradings for hair and wool type Dorpers of 9.8%-5.2% A1, 32.8%-39.9% A2, 40.2%-35.5% A3, 11.6%-11.9% A4, 5.1%-7.0% A5 and 0.5%-0.5% A6 respectively.

## 5.3   Overall discussion on Meatmaster production and reproduction traits

In animal production, reproduction and fecundity are considered the traits with the highest economical value and although not highly heritable according to Lasley (1972) the highest priority should constantly be assigned to reproduction ability (Campbell, 1983).

The difference in profitability of sheep breeds remains one of the most controversial issues among sheep producers. However, very little research has been done in this field on South African sheep breeds (Fourie and Cloete, 1993; Schoeman et al., 1995; Snyman et al., 2000; Cloete et al., 2004). The reason for this is that the full implication of breed comparisons is seldom taken into account (Snyman and Herselman, 2005). During the process of development of the new composite Meatmaster sheep breed most of the breeders observed differences between the resulting Meatmaster populations and their parent breeds. Where these differences were recorded it was not the intention to research a breed comparison but merely to report the observations which suggested improved performance by the Meatmaster sheep.

From the data recorded at the above mentioned localities an initial attempt can be made to characterize the production potential of the Meatmaster sheep breed.

### 5.3.1  Age at first lambing (A1stL)

In the Eastern Cape from 1999 to 2004 age at first lamb values for the Meatmasters gradually decreased from 17.8 months to 11.6 months in 2004. And between 2005 and 2008 the value varied from 12.0 to 13.3 months. This value is well established.

A phenotypic value of **13.0 months** for the trait of age at first lambing can be assigned to Meatmaster sheep in similar extensive production environments.

### 5.3.2  Lambing interval (ILP)

For the Eastern Cape where lambing seasons are controlled ILP varied from 328 to 398 days with a mean value of 350 days. Where lambing seasons are controlled such values do not represent a true value of the production potential of the ewe. The values recorded on the Highveld for ILP of Meatmaster sheep should represent a more acceptable value of the potential of Meatmaster sheep with respect to this production trait. Values recorded

on the Highveld in a free mating production environment varied between 252.13 and 273.03 days.

A phenotypic value of **273 days** for the trait of lambing interval can therefore be assigned to the Meatmaster sheep in a free mating system.

### 5.3.3  Lambing percentage (Ewes lambed / ewes mated)

Insufficient data were available to assign a specific phenotypic value for this trait but first indications are that the phenotypic value for Meatmasters can be expected to be equal to or better than the population mean value for mutton producing sheep. The lambing percentage value of 86% recorded for Meatmasters at Keetmanshoop in Namibia correlates well with published values for mutton producing sheep. The Meatmaster performed better than its parent breeds during the trail.

A phenotypic value of **86%** can be assigned to the Meatmaster breed for this trait.

### 5.3.4  Average litter size (fecundity %)

From the study by Snyman and Herselman (2005) it is also evident that fecundity values differed between breeds and for the same breeds between localities. From the recorded results it can be conclu-ded that fecundity values for Meatmaster sheep can be expected to be at least 6% higher than the population mean phenotypic value for Damara sheep and there is no indication that fecundity values for Meatmaster sheep could be expected to be lower than values recorded for Dorper sheep. From values recorded on the farm La Rochelle in the Eastern Cape Meatmaster fecundity varied from 106% to 132%. Keetmanshoop recorded 153% and on the Highveld life-time production of Meatmasters registered a fecundity value of 123%. It can therefore be concluded from the available data that Meatmaster sheep will have fecundity values ranging from 114% to 153% depending on the agro-ecosystem and selection objectives of different Meatmaster flock masters.

A phenotypic value of **120%** can be assigned to the Meatmaster for the trait of fecundity.

### 5.3.5   Survival percentage (lambs died from birth till weaning)

Meatmaster ewes in the Eastern Cape registered very high lamb survival percentages. Ewes that lambed between 3 and 6 times had a mean lamb survival rate of 97.2%.

A phenotypic value of **97%** for survival percentage can be accepted for mature Meatmaster ewes in comparable ecosystems.

### 5.3.6   Birth weight

Birth weight of Meatmaster lambs were recorded at Keetmanshoop as 3.35 kg which included more twins than single lambs. Birth weight varied between a minimum of 2.8 kg and a maximum of 4.4 kg.

Taking into account the fecundity value assigned to the Meatmaster a phenotypic value of **3.8 kg** can be assigned to the Meatmaster for this production trait.

*Fig 5.5   Meatmaster lambs are of average size and ewes have no lambing difficulties*

### 5.3.7   100-day (weaning) weight for ram and ewe lambs

Weaning weight for Meatmaster lambs generally fall between the weaning weight values for Damara lambs and those for Dorper lambs but are closer to the values for Dorper lambs than to the values obtained for Damara lambs. In harsh environmental conditions such as at Keetmanshoop, values for Meatmasters were very close to the Dorper weaning weight values. Meatmaster triplets were recorded at 26.99 kg and ram lambs at 27.09 kg. Between 1999 and 2009 Eastern Cape ram lambs registered mean 100-day weights of

between 22.6 kg and 29.3 kg and ewe lambs between 21.7 and 28.1 kg.

Taking into account that the phenotypic value for weaning weight is subject to the environmental and seasonal effects of variable conditions in production seasons an average value of **26.0 kg** can be assigned to the Meatmaster for this production trait.

### 5.3.8   270 day weight

270-day post weaning weights for Meatmaster lambs compare well to the values recorded for Dorper lambs of the same age. An overall mean value for single, twin and triplet Meatmaster lambs was recorded as 46.86 kg in the Eastern Cape. Single lambs at 48.01 kg, twins at 44.98 kg and triplets at 47.42 kg.

An average phenotypic value of **47.0 kg** can be assigned to Meatmasters for this production trait.

### 5.3.9   Live slaughter weight (LSW), cold carcass weight (CCW) and carcass grading (CG)

Limited data were available for the phenotypic evaluation of the slaughter traits for Meatmasters. Meatmaster lambs slaughtered at an average marketing weight of 38kg between 4 and 6 months of age should achieve acceptable gradings and dressing percentages with 80%-90% of A2 and A3 gradings. Collett (2007) reported that Meatmaster carcasses achieved 9th place amongst 59 groups entered for the National slaughter lamb competition held at Bethulie during 2006 and Russel (2010) reported from Australia that domestic markets there had no problem to absorb Meatmaster carcasses at the top of the prime lamb gradings. At Keetmanshoop all trail lambs were slaughtered at 212 days at a mean LSW of 35.95 kg of which 50% were twins. CCW recorded was 15.64 kg and gradings varied between A0 and A6.

Phenotypic values of **38 kg** LSW at **5 months** of age to yield a CCW of **17.5 kg** can be assigned to the Meatmaster for these production traits.

### 5.3.10   Mature ewe weight

Although limited data were available for a proper phenotypic evaluation of this trait, an estimate can be made from the value recorded for body weight on the Highveld and 270-day weights analysed. Meatmaster ewes were recorded on the Highveld with body weights varying from 46.0 kg to 64.0 kg with an average of 52.2 kg.

Snyman and Olivier, (2005) reported body weight for Dorper ewes at two localities as 57.8 kg and 55.4 kg respectively. Meatmaster ewes are expected to have higher body weights than Damara ewes and lower body weights compared to Dorper ewes. Ewes of average size have been proven to be the most functionally efficient breeding stock (Schoeman, 1996).

Meatmaster ewes adapt very well to the environment and the body weight of ewes is expected to vary from flock to flock in accordance with the effect of the local agro-ecosystem.

A phenotypic value of **52.2 kg** can be assigned to Meatmaster ewes for this trait.

*Fig 5.6    Two Meatmaster ewes with well reared triplet lambs*

### 5.3.11  Mature ram weight

Limited data were available for a proper phenotypic evaluation of this trait but an estimate can be made. Meatmaster rams were recorded on the Highveld with body weights varying from 50.0 kg to 66.0 kg with an average of 58.1 kg. (The age range of the rams was between 18 and 24 months). Older individual Meatmaster rams were recorded at weights of 80 kg, 95 kg and 105 kg on the Highveld farm. In the extensive sheep farming areas of the Eastern and Northern Cape mature ram weights are expected to be higher with an average in the range of 70 kg for 18 to 24 month old rams (personal observation).

A phenotypic value for 18 month old Meatmaster rams of **65 kg** can reasonably be assigned for this trait.

## 5.4  Conclusions

- The linear body measurements (LBM) recorded for Meatmaster sheep can be compared to measurements recorded for other sheep breeds and can be used by future researchers as reference values.
- Selection for specific traits by breeders may have an effect on the linear body measurements over a period of time. The effect of changes in the environment on linear body measurements of Meatmaster sheep can likewise be investigated in future research. The measurements for Meatmaster sheep recorded here is the first attempt to supplement the phenotypic description of the new Meatmaster sheep breed. The wealth of colours and colour patterns observed for Meatmaster sheep contributes to the economic potential of added value for exotic sheep skin products.
- Fecundity values recorded for Meatmasters varied between 114.0% and 153.5% at the three localities. Depending on selection programs practiced the Meatmaster can be developed into a very prolific breed.
- Meatmaster sheep have the proven ability to start production at an early age which contributes substantially to a shortening of the generation interval. The shorter generation interval of the Meatmaster contributes to the potential of faster genetic improvement of the breed.
- The easy-care Meatmaster sheep is ideally suited for production in extensive farming conditions with no supplementary feeding and provides an acceptable consumer product at low input costs.
- The ability of the Meatmaster to produce at short interlambing periods contributes to increased income generation, improved utilization of natural resources and enhancement of food security.
- Further research is recommended to supplement the currently available Meatmaster production data. Estimation of genetic and phenotypic correlations between lamb and ewe traits for Meatmaster sheep is recommended as soon as sufficient data are available on the INTERGIS system.

# ANYSBERG BOERDERY

## Meatmaster Breeders
### 33' 31 S | 20' 45 E

## BREEDING / TEEL VIR:

- Twins & Triplets / Meerlinge
- Greenfields or Veld / Aangeplante Wyding of Veld
- Hardiness / Gehardheid

- Motherhood / Moeder Eienskappe
- Weaning Weight / Speen Gewig
- Physical Size / Bou Vorm

**ANYSBERG BOERDERY • PAPKUILSFONTEIN • DISTRIK: LADISMITH, KAAP**

Enquiries/Navrae:  **Johan Saaiman**  028 551 1154 / 083 383 1000
                   **Pierre Rousseau**  011 323 2800 / 083 653 7310

Accommodation available on the farm / Akkommodasie beskikbaar op die plaas.

# Chapter 6   Agro-economics of the Meatmaster sheep breed

## 6.1   Introduction

Agro-economic reference material on the composite Meatmaster breed and other breeds is provided and the production potential of the Meatmaster at three different localities is evaluated.

Meat production is relatively inefficient in sheep and cattle. The high maintenance requirements of the breeding female contribute largely to this low efficiency (Dickerson, 1978). An improvement of both biological and economical efficiency therefore becomes increasingly important (Schoeman, 1996). Crossbreeding between highly productive, genetically divergent and adapted breeds may result in superior overall performance. The choice of breeds to be used in crossbreeding should match the level of inputs, which can vary substantially both within and amongst target areas and production systems (Philipsson et al., 2006).

The advantage of developing and using a synthetic breed is that it is easier to handle than systematic crossbreeding under practical conditions. Secondly, response to selection when using synthetic breeds is greater than that on parental breeds because of the increased genetic variation or increased additive genetic merit in the synthetics. Developing a synthetic breed, however, is a long-term process that needs more resources, a large number of animals, detailed recording and analytical facilities (Tibbo, 2006). No published data are available for the locally developed Meatmaster sheep breed. Data recorded on the Highveld in Gauteng, the farm La Rochelle in the Eastern Cape and the Gellap-Ost research station at Keetmanshoop in Namibia are used to prove that the Meatmaster sheep breed is economically more efficient in extensive agro-economical environments than its parent breeds, the Damara, Dorper, Ile de France and SA Mutton Merino. The Afrino, a dual purpose breed is included as an unrelated breed.

In a comparative study to compare the long term gross income of Afrino, Dorper and Merino sheep between 1956 and 1998 (Snyman et al., 1999) it was concluded that there was no significant difference in gross income between the three breeds. The study was done between March 1986 and March 1991 under veld conditions at Carnarvon and Tarka with ewe groups of 50 per breed. The conclusion in this case was dependent on the fact that over the time period the wool-meat price ratio was in the order of 2:1. The calculations over the five year period studied indicated however that the gross income of the Dorper group was the highest of the three breeds where a meat price of R12.00/kg, wool price of R17.40/kg and hide prices of R25.00 for Dorper hides and R5.00 for Merino hides were used. In this case the wool-meat price ratio was 1.45:1. The present (2010) wool-meat price ratio of R50.00/kg for wool (June 2010) and R38.00/kg (Colesberg abattoir, June 2010) for grade A2 lamb is even lower at 1.32:1. Snyman and Herselman (2005) reported that with wool and mutton prices used at the time of the specific study where three breeds (Afrino, Dorper and Merino) were compared at two different localities, Afrino sheep had the highest gross income per ewe, followed by Merino and Dorper ewes respectively. However, Merino sheep had the highest gross income per hectare at both localities. Afrino sheep performed intermediate, while Dorper sheep had the lowest gross income per hectare. The combination of low ewe body weights, high wool production and a relatively high reproductive rate, resulted in Merino sheep generating the highest income per hectare, compared to Afrino and Dorper sheep.

Wool and meat are two important commodities subject to price fluctuations dictated by consumer demand. It is therefore of strategic importance to maintain national flocks of sheep, of both wool and meat producing breeds, to support national economic flexibility in the case of major wool-meat price ratio changes.

There is however a growing international need for easier-managed sheep (Collins and Conington, 2010) based on the perception that easier-managed sheep, specifically of a hair or wool shedding type, will contribute to more profitable and productive sheep farming (Walker, 2010).

## 6.2 Evaluation of the agro-economic potential of Meatmaster sheep.

The following three data sets were used to assess the agro-economic value of Meatmaster sheep:

- Performance data were recorded between 1990 and 2009 on the Meatmaster population at the farm La Rochelle in the Venterstad district of the Eastern Cape as well as comparative data of the Damara and Dorper parent breeds, the SA Mutton Merino and the Afrino.
- Production data on the Gellapper population of Meatmaster sheep and its parent breeds were recorded at the Gellap-Ost research station in the Keetmanshoop district of Namibia.
- Production data were recorded between 1990 and 2006 for the Meatmaster population at the Schoongezicht farm on the Highveld in Gauteng, as well as its parent breeds, the Damara and Ile De France.

Phenotypic values for traits recorded and statistically analysed are discussed in Chapter 5. The data recorded at these localities of widely different ecosystems form a basis from which a preliminary assessment of the agro-economic value of the Meatmaster sheep breed can be performed.

The SM 2000 program developed by Herselman (2002) at the Grootfontein Agricultural Development Institute was applied to calculate the income per small stock unit (SSU), the gross margin per annum, per SSU and per hectare (ha). Stocking rates were calculated per calendar month as well as the comparative SSU's and large stock units (LSU). All calculations were based on a 1000 ha farm unit with a carrying capacity of 12ha/LSU and all calculations were performed based on actual LSU values obtained from the mean phenotypic values established at these localities.

Not all of the economically important traits were recorded at each locality but it is nevertheless possible to make fair estimations of these values. Where estimations of phenotypic values were made it was applied as constants between the breeds at a specific locality in order for the calculation to reflect the effect of the actually recorded traits.

Figure 6.1  Meatmaster

Figure 6.2  Damara

Figure 6.3  Ile de France

Figure 6.4  Dorper

## 6.3  Agro-economic results of the Meatmaster and parent breeds

### 6.3.1  La Rochelle Eastern Cape

The five sheep breeds at La Rochelle did not all run in the same herd but all ran on the same farm under fairly similar conditions. The Afrino and SA Mutton Merino are dual purpose breeds. The Meatmaster, Dorper and Damara are hair sheep breeds for meat production. The two dual purpose breeds have slightly better camps to graze but like all five the breeds they receive no licks, no supplementary feeding and no grazing on lands. Only salt is made available to all the breeds.

All the breeds are mated and rear their lambs under these entirely natural conditions. Even ewes with twin lambs are not separated but are left with all the other ewes to rear their lambs within the big herds under those extensive conditions. All the breeds, except the Damara, are mated to lamb only once a year either from the months of March/April or Sept/Oct. The first mating commences in October and only barren ewes are mated twice. The Damara breed is mated to lamb every 8 months in order to utilise the breed's potential to lamb at short inter lambing periods.

The input values used in the SM 2000 program for the calculations are listed in Table 6.1 for the 5 breeds at La Rochelle. Although 2 lambing seasons were required as input to the program in order to calculate the stocking rates ewes were only mated to lamb once a year. Therefore 50% were mated during October and the other half plus any barren ewes during April. For the Damara, mated to lamb 3 times in two years the first mating was in October and there after every 8 months.

Table 6.1  *Parameters and values applied for agro-economic calculations at La Rochelle Eastern Cape*

| Parameter | Afrino | Dorper | Meatmaster | Damara | SAMM |
|---|---|---|---|---|---|
| Replacement age (months) | 13 | 14 | 8 | 10 | 17 |
| Lambing seasons | 2 | 2 | 2 | 3 | 2 |
| Month mated | 10 | 10 | 10 | 10 | 10 |
| Lambing % | 113 | 115 | 114 | 109 | 121 |
| Birth weight (kg) | 3.8 | 3.7 | 3.55 | 3.42 | 3.8 |
| Weaning weight (kg) | 28.54 | 29.5 | 25.0 | 21.75 | 30.3 |
| 12 month W (kg) | 49 | 48 | 47 | 43 | 52 |
| Ewe weight (kg) | 55 | 55 | 52.2 | 46 | 60 |
| Mortality % till wean | 8.0 | 12.4 | 3.0 | 6.0 | 13.4 |
| Mortality % young | 4.0 | 4.0 | 2.0 | 3.0 | 4.0 |
| Mortality % mature | 2.0 | 2.0 | 1.0 | 1.0 | 2.0 |
| LSW (kg) | 42.0 | 38.0 | 38.0 | 38.0 | 48.0 |
| Sl Age (months) | 9 | 5 | 5 | 13 | 13 |
| Dressing % lambs | 42 | 49 | 47 | 43 | 42 |
| Dressing % sheep | 42 | 45 | 45 | 42 | 42 |
| Meat Price A (R/kg) | 38.00 | 38.00 | 38.00 | 38.00 | 38.00 |
| Meat Price B (R/kg) | 28.00 | 28.00 | 28.00 | 28.00 | 28.00 |
| Skin   (R/skin) | 10.00 | 30.00 | 30.00 | 30.00 | 10.00 |
| Head &T (R/u) | 15.00 | 15.00 | 15.00 | 15.00 | 15.00 |
| Old rams /head | 650.00 | 650.00 | 650.00 | 650.00 | 650.00 |
| Shearing Month | 9 | - | - | - | 9 |
| Wool (kg/ewe) | 2.01 | - | - | - | 2.51 |
| Wool price (R/kg) cy | 50.00 | - | - | - | 50.00 |
| Transport (R/SSU) | 8.00 | 8.00 | 8.00 | 8.00 | 8.00 |
| Marketing cost (%) | 5 | 5 | 5 | 5 | 5 |
| Shearing cost (R/SSU) | 7.55 | - | - | - | 7.55 |
| Wool marketing (%) | 4 | - | - | - | 4 |
| Clean yield (lambs) (%) | 62 | - | - | - | 62 |
| Clean yield (ewes) (%) | 67 | - | - | - | 67 |
| Ram replacement (R) | 4000 | 4000 | 4000 | 4000 | 4000 |
| Health costs (R/SSU) | 10.00 | 6.47 | 6.47 | 6.47 | 10.00 |
| Feed | 0 | 0 | 0 | 0 | 0 |

### 6.3.1.1  Afrino production at La Rochelle

A production system consisting of a 1000 ha farm, using actual LSU values for mutton sheep, a carrying capacity of 12 ha/LSU, stocking Afrino ewes in extensive farming conditions, with two lambing seasons (ewes first mated during October; only barren ewes mated twice), where 20% of the ewe flock is replaced annually with 13 months old young ewes, were applied. Lambs produced are slaughtered at the average age of 9 months (8-11 months) at a live weight of 42 kg, a dressing percentage of 42% and a prime lamb price of R38.00/kg (grade A2/A3), a slaughter price for older sheep of R28.00/kg at a dressing percentage of 42%. Skin price at R10.00/ head and trotters at R15.00 and a price obtainable for old rams at R650/head were used. Replacement cost of rams was taken

as R4000/head. Marketing costs were calculated as 5% of realisation and transportation costs of R8.00/head were applied. Birth weight of lambs of 3.8 kg, weaning weight of 28.54 kg, 12 month weight of 49.0 kg and adult ewe weight of 55 kg were included in the model. Furthermore, lambing percentage of 113%, survival rates of 92.0% for lambs, 96.0% for young sheep and 98.0% for adult sheep were used. No feeding costs except rock salt calculated at R10.00/head and health expenses of R10.00/head were allowed. Wool production parameters applied were  2.01 kg wool per annum for ewes with a clean yield of 67% and 62% for lambs at a wool price of R50.00/kg. The income and gross margins, livestock stocking rates and clip summary are summarised in Table 6.2 for Afrino sheep at La Rochelle.

Table 6.2  Income and gross margins for Afrino sheep at La Rochelle

|  | RAND | R/SSU | R/ha |
|---|---|---|---|
| **INCOME** | **245 166** | **441.30** | **245.17** |
| Animals | 211 959 | 381.53 | 211.96 |
| Products | 33 207 | 59.77 | 33.21 |
|  |  |  |  |
| **EXPENSES** | **30 622** | **55.12** | **30.62** |
| Animals | 491 | 0.88 | 0.49 |
| Feed | 0 | 0.00 | 0.00 |
| Shearing | 5 082 | 9.15 | 5.08 |
| Marketing | 13 938 | 25.09 | 13.94 |
| Health | 5 556 | 10.00 | 5.56 |
| Sundry | 5 556 | 10.00 | 5.56 |
|  |  |  |  |
| **GROSS MARGIN** | **214 544** | **386.18** | **214.54** |

### LIVESTOCK TABLE

| Month | NUMBER OF ANIMALS ON FARM | | | | | | Whethers | Rams | OTHER NUMBERS | | | STOCKING RATE | | |
|---|---|---|---|---|---|---|---|---|---|---|---|---|---|---|
|  | Offspring | | | Ewes | | |  |  | Sold | Bought | Replace | Head | LSU | SSU |
|  | Lamb | Weaner | 2-Tooth | Dry | Single | Twin |  |  |  |  |  |  |  |  |
| Jan | 0 | 191 | 0 | 305 | 0 | 0 | 0 | 6 | 0 | 0 | 0 | 502 | 73 | 488 |
| Feb | 0 | 190 | 0 | 305 | 0 | 0 | 0 | 6 | 0 | 0 | 0 | 501 | 75 | 497 |
| March | 173 | 189 | 0 | 176 | 85 | 43 | 0 | 6 | 0 | 0 | 0 | 673 | 76 | 508 |
| April | 170 | 189 | 0 | 177 | 86 | 44 | 0 | 6 | 28 | 0 | 31 | 671 | 86 | 573 |
| May | 166 | 157 | 0 | 177 | 86 | 44 | 0 | 6 | 0 | 0 | 0 | 636 | 91 | 606 |
| June | 163 | 157 | 0 | 177 | 86 | 44 | 0 | 6 | 126 | 0 | 0 | 632 | 99 | 661 |
| July | 0 | 191 | 0 | 305 | 0 | 0 | 0 | 6 | 0 | 0 | 0 | 502 | 73 | 488 |
| Aug | 0 | 190 | 0 | 305 | 0 | 0 | 0 | 6 | 0 | 0 | 0 | 501 | 75 | 497 |
| Sep | 173 | 189 | 0 | 176 | 85 | 43 | 0 | 6 | 0 | 0 | 0 | 673 | 76 | 508 |
| Oct | 170 | 189 | 0 | 177 | 86 | 44 | 0 | 6 | 28 | 0 | 31 | 672 | 86 | 573 |
| Nov | 166 | 157 | 0 | 177 | 86 | 44 | 0 | 6 | 0 | 0 | 0 | 636 | 91 | 606 |
| Dec | 163 | 157 | 0 | 177 | 86 | 44 | 0 | 6 | 126 | 0 | 0 | 632 | 99 | 661 |
| AVERAGE |  |  |  |  |  |  |  |  |  |  |  | 603 | 83 | 556 |

## CLIP SUMMARY

|  | Greasy wool (kg) | Clean wool (kg) | % |
|---|---|---|---|
| Lamb | 187 | 116 | 17 |
| Hogget | 186 | 124 | 19 |
| Main | 633 | 424 | 64 |
| **Total** | **1005** | **664** | **100** |

1005 kg wool was produced. Stocking rates in terms of heads on the farm varied from a minimum of 501 to a maximum of 673. Ewes contributed 57% to the LSU values and the offspring 41% with 2% from rams. The gross margin for income was R386.18/SSU and R214.54/ha.

## CONTRIBUTION OF ANIMAL CLASSES TO LSU'S

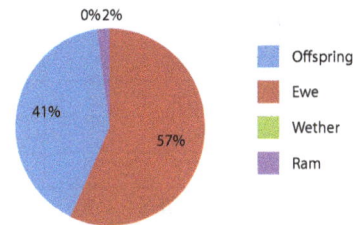

- Offspring
- Ewe
- Wether
- Ram

### 6.3.1.2 Dorper sheep production at La Rochelle

The production system as described for Afrino sheep, using the phenotypic and agro-economic values recorded for Dorper sheep at La Rochelle, as tabulated in Table 6.1 was applied for the Dorper sheep. The income and gross margins are summarised in Table 6.3.

Table 6.3   Income and gross margins for Dorper sheep at La Rochelle

|  | RAND | R/SSU | R/ha |
|---|---|---|---|
| **INCOME** | **261 239** | **470.23** | **261.24** |
| Animals | 261 239 | 470.23 | 261.24 |
| Products | 0 | 0.00 | 0.00 |
|  |  |  |  |
| **EXPENSES** | **27 017** | **48.63** | **27.02** |
| Animals | 570 | 1.03 | 0.57 |
| Feed | 0 | 0.00 | 0.00 |
| Shearing | 0 | 0.00 | 0.00 |
| Marketing | 15 336 | 27.60 | 15.34 |
| Health | 5 556 | 10.00 | 5.56 |
| Sundry | 5 556 | 10.00 | 5.56 |
|  |  |  |  |
| **GROSS MARGIN** | **234 221** | **421.60** | **234.22** |

Stocking rates in terms of heads on the farm varied from a minimum of 576 to a maximum of 638. Ewes contributed 66% to the LSU values and the offspring 32% with 2% from rams. The gross margin for income was R421.60/SSU and R234.22 /ha.

### 6.3.1.3 Meatmaster sheep production at La Rochelle

The production system as described for Afrino sheep, using the phenotypic and agro-economic values recorded for Meatmaster sheep at La Rochelle, as tabulated in Table 6.1 was applied for the Meatmaster sheep. The income and gross margins and livestock stocking rates are summarised in Table 6.4 for Meatmaster sheep at La Rochelle.

Table 6.4  Income and gross margins for Meatmaster sheep at La Rochelle

|  | RAND | R/SSU | R/ha |
|---|---|---|---|
| **INCOME** | **320 626** | **577.13** | **320.63** |
| Animals | 320 626 | 577.13 | 320.63 |
| Products | 0 | 0.00 | 0.00 |
|  |  |  |  |
| **EXPENSES** | **30 737** | **55.33** | **30.74** |
| Animals | 652 | 1.17 | 0.65 |
| Feed | 0 | 0.00 | 0.00 |
| Shearing | 0 | 0.00 | 0.00 |
| Marketing | 18 975 | 34.15 | 18.97 |
| Health | 5 556 | 10.00 | 5.56 |
| Sundry | 5 556 | 10.00 | 5.56 |
|  |  |  |  |
| **GROSS MARGIN** | **289 889** | **521.80** | **289.89** |

## LIVESTOCK TABLE

| | NUMBER OF ANIMALS ON FARM | | | | | | Whethers | Rams | OTHER NUMBERS | | | STOCKING RATE | | |
|---|---|---|---|---|---|---|---|---|---|---|---|---|---|---|
| | Offspring | | | Ewes | | | | | Sold | Bought | Replace | Head | LSU | SSU |
| Month | Lamb | Weaner | 2-Tooth | Dry | Single | Twin | | | | | | | | |
| Jan | 0 | 225 | 0 | 407 | 0 | 0 | 0 | 8 | 0 | 0 | 0 | 640 | 88 | 585 |
| Feb | 0 | 225 | 0 | 406 | 0 | 0 | 0 | 8 | 184 | 0 | 0 | 639 | 90 | 598 |
| March | 232 | 41 | 0 | 234 | 113 | 59 | 0 | 8 | 0 | 0 | 0 | 687 | 68 | 455 |
| April | 230 | 41 | 0 | 234 | 112 | 59 | 0 | 8 | 0 | 0 | 0 | 685 | 78 | 519 |
| May | 229 | 41 | 0 | 235 | 113 | 60 | 0 | 8 | 39 | 0 | 41 | 685 | 87 | 581 |
| June | 227 | 0 | 0 | 235 | 113 | 60 | 0 | 8 | 0 | 0 | 0 | 642 | 89 | 596 |
| July | 0 | 225 | 0 | 407 | 0 | 0 | 0 | 8 | 0 | 0 | 0 | 640 | 88 | 584 |
| Aug | 0 | 225 | 0 | 406 | 0 | 0 | 0 | 8 | 184 | 0 | 0 | 639 | 90 | 598 |
| Sep | 232 | 41 | 0 | 234 | 113 | 59 | 0 | 8 | 0 | 0 | 0 | 687 | 68 | 455 |
| Oct | 230 | 41 | 0 | 234 | 112 | 59 | 0 | 8 | 0 | 0 | 0 | 685 | 78 | 519 |
| Nov | 229 | 41 | 0 | 235 | 113 | 60 | 0 | 8 | 39 | 0 | 41 | 685 | 87 | 581 |
| Dec | 227 | 0 | 0 | 235 | 113 | 60 | 0 | 8 | 0 | 0 | 0 | 642 | 89 | 596 |
| AVERAGE | | | | | | | | | | | | 663 | 83 | 556 |

## CONTRIBUTION OF ANIMAL CLASSES TO LSU'S

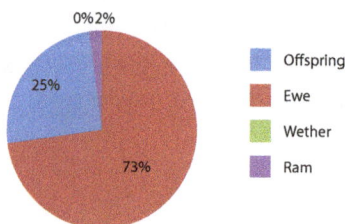

Stocking rates in terms of heads on the farm varied from a minimum of 639 to a maximum of 687. Ewes contributed 73% to the LSU values and the offspring 25% with 2% from rams. The gross margin for income was R521.80/SSU and R289.89/ha.

### 6.3.1.4  *Damara sheep production at La Rochelle*

The Damara sheep were mated to lamb three times in two years, using the phenotypic and agro-economic values recorded for Damara sheep at La Rochelle, as tabulated in Table 6.1, the income and gross margins and livestock stocking rates were calculated and are summarised in Table 6.5.

Table 6.5  Income and gross margins for Damara sheep at La Rochelle

|  | RAND | R/SSU | R/ha |
|---|---|---|---|
| **INCOME** | **262 413** | **472.34** | **262.41** |
| Animals | 262 413 | 472.34 | 262.41 |
| Products | 0 | 0.00 | 0.00 |
|  |  |  |  |
| **EXPENSES** | **27 455** | **49.42** | **27.45** |
| Animals | 429 | 0.77 | 0.43 |
| Feed | 0 | 0.00 | 0.00 |
| Shearing | 0 | 0.00 | 0.00 |
| Marketing | 15 915 | 28.65 | 15.91 |
| Health | 5 556 | 10.00 | 5.56 |
| Sundry | 5 556 | 10.00 | 5.56 |
|  |  |  |  |
| **GROSS MARGIN** | **234 958** | **422.93** | **234.96** |

## LIVESTOCK TABLE

| Month | Lamb | Weaner | 2-Tooth | Dry | Single | Twin | Whethers | Rams | Sold | Bought | Replace | Head | LSU | SSU |
|---|---|---|---|---|---|---|---|---|---|---|---|---|---|---|
| Jan | 0 | 273 | 0 | 267 | 0 | 0 | 0 | 5 | 0 | 0 | 0 | 545 | 69 | 458 |
| Feb | 0 | 272 | 0 | 267 | 0 | 0 | 0 | 5 | 0 | 0 | 0 | 544 | 71 | 474 |
| March | 292 | 272 | 0 | 44 | 155 | 68 | 0 | 5 | 0 | 0 | 0 | 836 | 73 | 489 |
| April | 288 | 271 | 0 | 44 | 155 | 68 | 0 | 5 | 0 | 0 | 0 | 830 | 83 | 554 |
| May | 283 | 270 | 0 | 44 | 156 | 68 | 0 | 5 | 34 | 0 | 36 | 827 | 94 | 628 |
| June | 279 | 234 | 0 | 44 | 156 | 68 | 0 | 5 | 0 | 0 | 0 | 786 | 99 | 659 |
| July | 0 | 508 | 0 | 267 | 0 | 0 | 0 | 5 | 0 | 0 | 0 | 781 | 87 | 582 |
| Aug | 0 | 507 | 0 | 267 | 0 | 0 | 0 | 5 | 233 | 0 | 0 | 779 | 90 | 599 |
| Sep | 0 | 273 | 0 | 267 | 0 | 0 | 0 | 5 | 0 | 0 | 0 | 545 | 69 | 458 |
| Oct | 0 | 272 | 0 | 267 | 0 | 0 | 0 | 5 | 0 | 0 | 0 | 544 | 71 | 474 |
| Nov | 292 | 272 | 0 | 44 | 155 | 68 | 0 | 5 | 0 | 0 | 0 | 836 | 73 | 489 |
| Dec | 288 | 271 | 0 | 44 | 155 | 68 | 0 | 5 | 0 | 0 | 0 | 830 | 83 | 554 |
| Jan* | 283 | 270 | 0 | 44 | 156 | 68 | 0 | 5 | 34 | 0 | 36 | 827 | 94 | 628 |
| Feb* | 279 | 234 | 0 | 44 | 156 | 68 | 0 | 5 | 0 | 0 | 0 | 786 | 99 | 659 |
| Mar* | 0 | 508 | 0 | 267 | 0 | 0 | 0 | 5 | 0 | 0 | 0 | 781 | 87 | 582 |
| April* | 0 | 507 | 0 | 267 | 0 | 0 | 0 | 5 | 233 | 0 | 0 | 779 | 90 | 599 |
| May* | 0 | 273 | 0 | 267 | 0 | 0 | 0 | 5 | 0 | 0 | 0 | 545 | 69 | 458 |
| June* | 0 | 272 | 0 | 267 | 0 | 0 | 0 | 5 | 0 | 0 | 0 | 544 | 71 | 474 |
| July* | 292 | 272 | 0 | 44 | 155 | 68 | 0 | 5 | 0 | 0 | 0 | 836 | 73 | 489 |
| Aug* | 288 | 271 | 0 | 44 | 155 | 68 | 0 | 5 | 0 | 0 | 0 | 830 | 83 | 554 |
| Sep* | 283 | 270 | 0 | 44 | 156 | 68 | 0 | 5 | 34 | 0 | 36 | 827 | 94 | 628 |
| Oct* | 279 | 234 | 0 | 44 | 156 | 68 | 0 | 5 | 0 | 0 | 0 | 786 | 99 | 659 |
| Nov* | 0 | 508 | 0 | 267 | 0 | 0 | 0 | 5 | 0 | 0 | 0 | 781 | 87 | 582 |
| Dec* | 0 | 507 | 0 | 267 | 0 | 0 | 0 | 5 | 233 | 0 | 0 | 779 | 90 | 599 |
| AVERAGE |  |  |  |  |  |  |  |  |  |  |  | 741 | 83 | 556 |

*Only applicable if a system of three matings in two years is practiced*

Stocking rates in terms of heads on the farm varied from a minimum of 544 to a maximum of 836. Ewes contributed 43% to the LSU values and the offspring 56% with 1% from rams. The gross margin for income was R422.93/SSU and R234.96/ha.

Noteworthy is that the gross income margin per SSU as well as the gross margin per ha for the Damara equals the respective income values for Dorper sheep if the potential of the Damara breed to reproduce at shorter lambing intervals is utilized

and provided that there is no meat price discrimination against the Damara sheep.

### 6.3.1.5  SA Mutton Merino production at La Rochelle

The production system as described for Afrino sheep, using the phenotypic and agro-economic values recorded for SA Mutton Merino sheep at La Rochelle, as tabulated in Table 6.1 was applied for the SA Mutton Merino sheep. The income and gross margins are summarised in Table 6.6.

Table 6.6   Income and gross margins for SA Mutton Merino at La Rochelle

|  | RAND | R/SSU | R/ha |
|---|---|---|---|
| **INCOME** | **221 759** | **399.17** | **221.76** |
| Animals | 183 724 | 330.70 | 183.72 |
| Products | 38 036 | 68.46 | 38.04 |
|  |  |  |  |
| **EXPENSES** | **28 490** | **51.28** | **28.49** |
| Animals | 379 | 0.68 | 0.38 |
| Feed | 0 | 0.00 | 0.00 |
| Shearing | 4 749 | 8.55 | 4.75 |
| Marketing | 12 250 | 22.05 | 12.25 |
| Health | 5 556 | 10.00 | 5.56 |
| Sundry | 5 556 | 10.00 | 5.56 |
|  |  |  |  |
| **GROSS MARGIN** | **193 270** | **347.89** | **193.27** |

**CLIP SUMMARY**

|  | Greasy wool (kg) | Clean wool (kg) | % |
|---|---|---|---|
| Lamb | 186 | 115 | 15 |
| Hogget | 316 | 212 | 28 |
| Main | 647 | 434 | 57 |
| **Total** | **1149** | **761** | **100** |

Stocking rates in terms of heads on the farm varied from a minimum of 509 to a maximum of 629. Ewes contributed 48% to the LSU values and the offspring 51% with 1% from rams. The gross margin for income was R347.89/SSU and R193.27/ha. 1149kg wool was produced.

For ease of comparison, the income and gross margins for the five breeds at La Rochelle are summarised in table 6.7.

Table 6.7.   Summary of income and gross margins for the five breeds at La Rochelle

|  | Afrino | Dorper | Meatmaster | Damara | SAMM |
|---|---|---|---|---|---|
| Income | 245 166.00 | 261 239.00 | 320 626.00 | 262 413.00 | 221 759.00 |
| Income/SSU | 441.00 | 470.23 | 577.13 | 472.34 | 399.17 |
| Income/ha | 245.17 | 261.24 | 320.63 | 262.41 | 221.76 |
|  |  |  |  |  |  |
| Gross margin | 214 544.00 | 234 221.00 | 289 889.00 | 234 958.00 | 193 270.00 |
| Gross margin/SSU | 386.18 | 421.60 | 521.80 | 422.93 | 347.89 |
| Gross margin/ha | 214.54 | 234.22 | 289.89 | 234.96 | 193.27 |

Figure 6.5 provides a graphical summary of the gross margin income values per SSU for the 5 breeds at La Rochelle. The Meatmaster gross margin income is substantially above all of the other four breeds, with a premium for Meatmasters of R100/SSU to R174/SSU at current (2010) wool and meat prices.

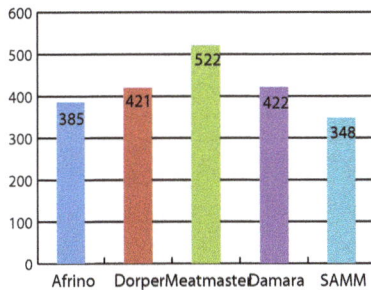

*Fig 6.5    Gross margin per SSU for the five breeds at La Rochelle*

### 6.3.2  Keetmanshoop

Flocks of Dorper, Damara and Meatmaster (Gellapper) sheep are kept in extensive farming conditions at the Gellap-Ost research station at Keetmanshoop in Namibia. Trials were done to establish the production and economic viability of the Gellapper line of Meatmaster sheep during the 2003/2004 season. A group of about 50 ewes from each breed was mated to Meatmaster (Ge), Damara and Dorper rams and the results recorded. The trail was only run for one season and all the progeny were slaughtered at the same time although it was obvious that many of the twin lambs lacked market readiness. The agro-economic calculation is reviewed by calculation with the SM 2000 program and current meat prices in order to be able to compare results calculated on the same basis for the three localities.

*Table 6.8   Parameters and values applied for agro-economic calculations at Keetmanshoop*

| Parameter | Dorper | Meatmaster | Damara |
|---|---|---|---|
| Replacement Age (Months) | 12 | 12 | 12 |
| Times mated | 1 | 1 | 1 |
| Month mated | 10 | 10 | 10 |
| Lambing % | 80.0 | 132.0 | 80.4 |
| Birth weight (kg) | 3.7 | 3.35 | 3.42 |
| Weaning weight (kg) | 25.3 | 24.9 | 18.6 |
| 12 month W (kg) | 50 | 48.5 | 45 |
| Ewe weight (kg) | 57 | 55 | 48 |
| Mortality % till wean | 14.0 | 18.0 | 17.0 |
| Mortality % young | 4.0 | 4.0 | 4.0 |
| Mortality % mature | 2.0 | 2.0 | 2.0 |
| LSW (kg) | 38.0 | 38.0 | 38.0 |
| SI Age (months) | 7 | 7 | 7 |
| Dressing %lambs | 43.71 | 43.51 | 42.7 |
| Dressing % sheep | 42 | 42 | 42 |
| Meat Price A (R/kg) | 38.00 | 38.00 | 38.00 |
| Meat Price B (R/kg) | 28.00 | 28.00 | 28.00 |
| Skin   (R/skin) | 30.00 | 30.00 | 30.00 |
| Head &T (R/u) | 15.00 | 15.00 | 15.00 |
| Old rams /head | 650.00 | 650.00 | 650.00 |
| Shearing Month | - | - | - |
| Wool (kg/ewe) | - | - | - |
| Wool price (R/kg) c y | - | - | - |
| Transport (R/SSU)  8.00 | 8.00 | 8.00 | |
| Marketing cost (%) | 5 | 5 | 5 |
| Shearing cost (R/SSU) | - | - | - |
| Wool marketing (%) | - | - | - |
| Clean yield (lambs) | - | - | - |
| Clean yield (Ewes) | - | - | - |
| Ram replacement (R) | 4000 | 4000 | 4000 |
| Health costs (R/SSU) | 6.47 | 6.47 | 6.47 |
| Feed | 0 | 0 | 0 |

### 6.3.2.1 Keetmanshoop Meatmaster (Ge) Sheep

Calculations were performed with the SM 2000 program based on a production system consisting of a 1000 ha farm with a carrying capacity of 12ha/LSU, stocking Meatmaster (Gellapper) ewes in extensive farming conditions, to lamb once per year based on the production parameters as listed in Table 6.8 to determine the gross margin and stocking rates in accordance with the phenotypic values measured for productive traits of the three breeds recorded at Keetmanshoop.

The income and gross margins are summarised in Table 6.9.

Table 6.9   Income and gross margins for Meatmasters (Ge) at Keetmanshoop

|  | RAND | R/SSU | R/ha |
|---|---|---|---|
| **INCOME** | **229 482** | **413.07** | **229.48** |
| Animals | 229 482 | 413.07 | 229.48 |
| Products | 0 | 0.00 | 0.00 |
|  |  |  |  |
| **EXPENSES** | **25 471** | **45.85** | **25.47** |
| Animals | 540 | 0.97 | 0.54 |
| Feed | 0 | 0.00 | 0.00 |
| Shearing | 0 | 0.00 | 0.00 |
| Marketing | 13 819 | 24.87 | 13.82 |
| Health | 5 556 | 10.00 | 5.56 |
| Sundry | 5 556 | 10.00 | 5.56 |
|  |  |  |  |
| **GROSS MARGIN** | **204 011** | **367.22** | **204.01** |

## LIVESTOCK TABLE

| Month | NUMBER OF ANIMALS ON FARM | | | | | | Whethers | Rams | OTHER NUMBERS | | | STOCKING RATE | | |
|---|---|---|---|---|---|---|---|---|---|---|---|---|---|---|
|  | Offspring | | | Ewes | | | | | Sold | Bought | Replace | Head | LSU | SSU |
|  | Lamb | Weaner | 2-Tooth | Dry | Single | Twin | | | | | | | | |
| Jan | 0 | 68 | 0 | 332 | 0 | 0 | 0 | 7 | 0 | 0 | 0 | 407 | 66 | 437 |
| Feb | 0 | 68 | 0 | 332 | 0 | 0 | 0 | 7 | 0 | 0 | 0 | 406 | 66 | 441 |
| March | 446 | 68 | 0 | 42 | 147 | 150 | 0 | 7 | 61 | 0 | 68 | 858 | 66 | 440 |
| April | 424 | 0 | 0 | 41 | 146 | 149 | 0 | 7 | 0 | 0 | 0 | 768 | 75 | 497 |
| May | 404 | 0 | 0 | 41 | 146 | 149 | 0 | 7 | 0 | 0 | 0 | 747 | 89 | 594 |
| June | 384 | 0 | 0 | 41 | 146 | 149 | 0 | 7 | 0 | 0 | 0 | 727 | 102 | 680 |
| July | 0 | 366 | 0 | 335 | 0 | 0 | 0 | 7 | 0 | 0 | 0 | 708 | 96 | 643 |
| Aug | 0 | 364 | 0 | 335 | 0 | 0 | 0 | 7 | 0 | 0 | 0 | 706 | 100 | 667 |
| Sep | 0 | 363 | 0 | 334 | 0 | 0 | 0 | 7 | 0 | 0 | 0 | 704 | 104 | 691 |
| Oct | 0 | 362 | 0 | 334 | 0 | 0 | 0 | 7 | 293 | 0 | 0 | 702 | 107 | 715 |
| Nov | 0 | 68 | 0 | 333 | 0 | 0 | 0 | 7 | 0 | 0 | 0 | 408 | 64 | 429 |
| Dec | 0 | 68 | 0 | 333 | 0 | 0 | 0 | 7 | 0 | 0 | 0 | 408 | 65 | 433 |
| AVERAGE |  |  |  |  |  |  |  |  |  |  |  | 629 | 83 | 556 |

## CONTRIBUTION OF ANIMAL CLASSES TO LSU'S

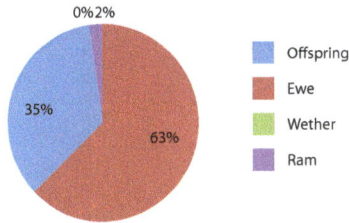

Legend: Offspring, Ewe, Wether, Ram

Stocking rates in terms of heads on the farm varied from a minimum of 406 to a maximum of 858. Ewes contributed 63% to the LSU values and the offspring 35% with 2% from rams. The gross margin for income was R367.22/SSU and R204.01/ha.

### 6.3.2.2 Keetmanshoop Dorper Sheep

The production system as described for Meatmaster sheep, using the phenotypic and agro-economic values recorded for Dorper sheep at Keetmanshoop, as tabulated in Table 6.8 was applied for the Dorper sheep. The income and gross margins are summarised in Table 6.10 for Dorper sheep at Keetmanshoop.

Table 6.10   Income and gross margins for Dorper sheep at Keetmanshoop

|  | RAND | R/SSU | R/ha |
|---|---|---|---|
| **INCOME** | **166 840** | **300.31** | **166.84** |
| Animals | 166 840 | 300.31 | 166.84 |
| Products | 0 | 0.00 | 0.00 |
|  |  |  |  |
| **EXPENSES** | **21 497** | **38.69** | **21.50** |
| Animals | 603 | 1.09 | 0.60 |
| Feed | 0 | 0.00 | 0.00 |
| Shearing | 0 | 0.00 | 0.00 |
| Marketing | 9 782 | 17.61 | 9.78 |
| Health | 5 556 | 10.00 | 5.56 |
| Sundry | 5 556 | 10.00 | 5.56 |
|  |  |  |  |
| **GROSS MARGIN** | **145 343** | **261.62** | **145.34** |

Stocking rates in terms of heads on the farm varied from a minimum of 453 to a maximum of 761. Ewes contributed 72% to the LSU values and the offspring 26% with 2% from rams. The gross margin for income was R261.62/SSU and R145.34/ha.

### 6.3.2.3 Keetmanshoop Damara Sheep

The production system as described for Meatmaster sheep, using the phenotypic and agro-economic values recorded for Damara sheep at Keetmanshoop, as tabulated in Table 6.8 was applied for the Damara sheep. The income and gross margins are summarised in Table 6.11 for Damara sheep at Keetmanshoop.

*Table 6.11   Income and gross margins for Damara sheep at Keetmanshoop*

|  | RAND | R/SSU | R/ha |
|---|---|---|---|
| **INCOME** | **157 041** | **282.67** | **157.04** |
| Animals | 157 041 | 282.67 | 157.04 |
| Products | 0 | 0.00 | 0.00 |
|  |  |  |  |
| **EXPENSES** | **21 320** | **38.38** | **21.32** |
| Animals | 717 | 1.29 | 0.72 |
| Feed | 0 | 0.00 | 0.00 |
| Shearing | 0 | 0.00 | 0.00 |
| Marketing | 9 491 | 17.08 | 9.49 |
| Health | 5 556 | 10.00 | 5.56 |
| Sundry | 5 556 | 10.00 | 5.56 |
|  |  |  |  |
| **GROSS MARGIN** | **135 721** | **244.30** | **135.72** |

Stocking rates in terms of heads on the farm varied from a minimum of 539 to a maximum of 907. Ewes contributed 75% to the LSU values and the offspring 23% with 2% from rams. The gross margin for income was R244.30/SSU and R135.72/ha.

For ease of comparison, the income and gross margins for the three breeds at Keetmanshoop are summarised in Table 6.12.

*Table 6.12  Summary of income and gross margins for the three breeds at Keetmanshoop*

|  | Dorper | Meatmaster | Damara |
|---|---|---|---|
| Income | 166 840 | 229 482 | 157 041 |
| Income/SSU | 300.31 | 413.07 | 282.67 |
| Income/ha | 166.84 | 229.48 | 157.04 |
|  |  |  |  |
| Gross margin | 145 343 | 204 011 | 135 721 |
| Gross margin/SSU | 261.62 | 367.22 | 244.30 |
| Gross margin/ha | 145.34 | 204.01 | 135.72 |

Figure 6.6 provides a graphical summary of the gross margin income values per SSU for the three breeds at Keetmanshoop.

*Fig 6.6  Gross margin per SSU for the three breeds at Keetmanshoop*

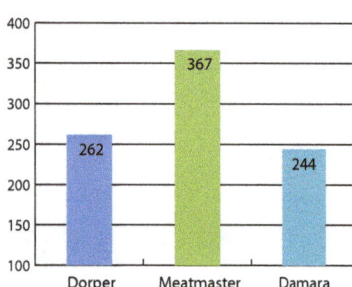

### 6.3.3  Results obtained on the farm Schoongezicht on the Highveld in Gauteng

Data recorded between 1990 and 2006 were included in table 6.13 as applicable parameters to use in the SM 2000 program. Over a 12 month period (2004 season) the Meatmaster stud flock at Schoongezicht was analysed to determine the productivity potential per annum of fully grown Meatmaster ewes of 36 to 60 months old (Table 5.1). The Average Lamb Production per annum (ALP) was found to be in excess of 200%, illustrating the exceptional production potential associated with the Meatmaster breed (Peters, 2006). The flocks at Schoongezicht were run in a free mating system since the onset of the breed creation programme. The three breeds were managed separately but the same conditions applied. The SM 2000 programme does not provide a specific option for a free mating system and it was therefore necessary to calculate the ALP per annum from the fecundity and inter lambing periods established. The algorithm to calculate the average lamb production% per annum (ALP% p.a.) is

$$\text{ALP\% p.a.} = \frac{365 \text{ (days)}}{\text{Interlambing Interval (days)}} \times \text{fecundity} \times 100\%$$

The SM 2000 programme option of 3 matings in two years was applied for the Highveld breed agro-economic calculations. From the ALP% values determined for the breeds, corrected lambing percentage values were calculated. (for example ALP% = 150% p.a., for two years production it would be 300/3% = 100% p.a., as lambing% input for the SM 2000 program to lamb 3 times in two years).

*Table 6.13  Parameters and values applied for agro-economic calculations at Schoongezicht on the Highveld*

| Parameter | Ile De France | Meatmaster | Damara |
|---|---|---|---|
| Replacement Age (Months) | 18.2 | 9.6 | 11.7 |
| Times mated | 3 | 3 | 3 |
| Month mated | 10 | 10 | 10 |
| Lambing % | 91.41 | 109.63 | 99.71 |
| Birth weight (kg) | 4.0 | 3.35 | 3.1 |
| Weaning weight (kg) | 29.9 | 24.9 | 18.6 |
| 12 month W (kg) | 52 | 47 | 43 |
| Ewe weight (kg) | 60 | 52.2 | 46 |
| Mortality % till wean | 14.0 | 14.0 | 14.0 |
| Mortality % young | 4.0 | 4.0 | 4.0 |
| Mortality % mature | 2.0 | 2.0 | 2.0 |
| LSW (kg) | 52.0 | 47.0 | 43.0 |
| SI Age (months) | 12 | 12 | 15 |
| Dressing % lambs | 43.5 | 43.5 | 43.5 |
| Dressing % sheep | 42 | 42 | 42 |
| Meat Price A (R/kg) | 38.00 | 38.00 | 38.00 |
| Meat Price B (R/kg) | 28.00 | 28.00 | 28.00 |
| Skin  (R/skin) | 10.00 | 30.00 | 30.00 |
| Head &T (R/u) | 15.00 | 15.00 | 15.00 |
| Old rams /head | 650.00 | 650.00 | 650.00 |
| Shearing Month | 9 | - | - |
| Wool (kg/ewe) | 2.5 | - | - |
| Wool price (R/kg) c y | 50.00 | - | - |
| Transport (R/SSU) | 8.00 | 8.00 | 8.00 |
| Marketing cost (%) | 5 | 5 | 5 |
| Shearing cost (R/SSU) | 7.75 | - | - |
| Wool marketing (%) | 4 | - | - |
| Clean yield (lambs) | 62 | - | - |
| Clean yield (Ewes) | 67 | - | - |
| Ram replacement (R) | 4000 | 4000 | 4000 |
| Health costs (R/SSU) | 10.00 | 6.47 | 6.47 |
| Feed | 0 | 0 | 0 |

### 6.3.3.1  Highveld Damara sheep

Calculations were performed with the SM 2000 program based on a production system consisting of a 1000 ha farm with a carrying capacity of 12ha/LSU, stocking ewes in extensive farming conditions, to lamb 3 times in two years based on the production parameters as listed in Table 6.13, to determine the gross margin income and stocking rates in accordance with the phenotypic values measured for productive traits of the three breeds at Schoongezicht on the Highveld.

The income and gross margins are summarised in Table 6.14 for Damara sheep on the Highveld.

*Table 6.14  Income and gross margins for Damara sheep on the Highveld*

|  | RAND | R/SSU | R/ha |
|---|---|---|---|
| **INCOME** | **237 662** | **427.79** | **237.66** |
| Animals | 237 662 | 427.79 | 237.66 |
| Products | 0 | 0.00 | 0.00 |
|  |  |  |  |
| **EXPENSES** | **26 225** | **47.20** | **26.22** |
| Animals | 449 | 0.81 | 0.45 |
| Feed | 0 | 0.00 | 0.00 |
| Shearing | 0 | 0.00 | 0.00 |
| Marketing | 14 665 | 26.40 | 14.66 |
| Health | 5 556 | 10.00 | 5.56 |
| Sundry | 5 556 | 10.00 | 5.56 |
|  |  |  |  |
| **GROSS MARGIN** | **211 437** | **380.59** | **211.44** |

Stocking rates in terms of heads on the farm varied from a minimum of 738 to a maximum of 791. Ewes contributed 43% to the LSU values and the offspring 56% with 1% from rams. The gross margin for income was R380.59/SSU and R211.44/ha.

### 6.3.3.2  Highveld Meatmaster sheep

The production system as described for Damara sheep, using the phenotypic and agro-economic values recorded for Meatmaster sheep at Schoongezicht on the Highveld, as tabulated in Table 6.13 was applied for the Meatmaster sheep. The income and gross margins and livestock stocking rates are summarised in Table 6.15 for Meatmaster sheep at Schoongezicht on the Highveld.

*Table 6.15  Income and gross margins for Meatmaster sheep on the Highveld*

|  | RAND | R/SSU | R/ha |
|---|---|---|---|
| **INCOME** | **302 609** | **544.70** | **302.61** |
| Animals | 302 609 | 544.70 | 302.61 |
| Products | 0 | 0.00 | 0.00 |
|  |  |  |  |
| **EXPENSES** | **29 509** | **53.12** | **29.51** |
| Animals | 415 | 0.75 | 0.41 |
| Feed | 0 | 0.00 | 0.00 |
| Shearing | 0 | 0.00 | 0.00 |
| Marketing | 17 983 | 32.37 | 17.98 |
| Health | 5 556 | 10.00 | 5.56 |
| Sundry | 5 556 | 10.00 | 5.56 |
|  |  |  |  |
| **GROSS MARGIN** | **273 101** | **491.58** | **273.10** |

**LIVESTOCK TABLE**

| | NUMBER OF ANIMALS ON FARM | | | | | | | | OTHER NUMBERS | | | STOCKING RATE | | |
|---|---|---|---|---|---|---|---|---|---|---|---|---|---|---|
| | Offspring | | | Ewes | | | Whethers | Rams | Sold | Bought | Replace | Head | LSU | SSU |
| Month | Lamb | Weaner | 2-Tooth | Dry | Single | Twin | | | | | | | | |
| Jan | 0 | 243 | 0 | 248 | 0 | 0 | 0 | 5 | 0 | 0 | 0 | 496 | 70 | 466 |
| Feb | 0 | 242 | 0 | 248 | 0 | 0 | 0 | 5 | 0 | 0 | 0 | 495 | 72 | 480 |
| March | 284 | 241 | 0 | 40 | 143 | 64 | 0 | 5 | 0 | 0 | 0 | 778 | 74 | 494 |
| April | 274 | 240 | 0 | 40 | 143 | 64 | 0 | 5 | 0 | 0 | 0 | 766 | 87 | 579 |
| May | 263 | 239 | 0 | 40 | 143 | 64 | 0 | 5 | 0 | 0 | 0 | 755 | 99 | 658 |
| June | 254 | 239 | 0 | 40 | 143 | 64 | 0 | 5 | 0 | 0 | 0 | 744 | 109 | 729 |
| July | 0 | 482 | 0 | 246 | 0 | 0 | 0 | 5 | 238 | 0 | 0 | 733 | 92 | 615 |
| Aug | 0 | 243 | 0 | 246 | 0 | 0 | 0 | 5 | 0 | 0 | 0 | 494 | 67 | 449 |
| Sep | 0 | 243 | 0 | 245 | 0 | 0 | 0 | 5 | 0 | 0 | 0 | 493 | 69 | 463 |
| Oct | 0 | 242 | 0 | 245 | 0 | 0 | 0 | 5 | 0 | 0 | 0 | 492 | 72 | 477 |
| Nov | 284 | 241 | 0 | 40 | 141 | 63 | 0 | 5 | 0 | 0 | 0 | 774 | 74 | 491 |
| Dec | 274 | 240 | 0 | 40 | 141 | 63 | 0 | 5 | 0 | 0 | 0 | 763 | 86 | 576 |
| Jan* | 263 | 239 | 0 | 40 | 141 | 63 | 0 | 5 | 0 | 0 | 0 | 751 | 98 | 654 |
| Feb* | 254 | 239 | 0 | 40 | 141 | 63 | 0 | 5 | 0 | 0 | 0 | 740 | 109 | 726 |
| Mar* | 0 | 482 | 0 | 243 | 0 | 0 | 0 | 5 | 238 | 0 | 0 | 730 | 92 | 612 |
| April* | 0 | 243 | 0 | 242 | 0 | 0 | 0 | 5 | 0 | 0 | 0 | 491 | 67 | 445 |
| May* | 0 | 243 | 0 | 242 | 0 | 0 | 0 | 5 | 0 | 0 | 0 | 490 | 69 | 460 |
| June* | 0 | 242 | 0 | 241 | 0 | 0 | 0 | 5 | 0 | 0 | 0 | 488 | 71 | 474 |
| July* | 284 | 241 | 0 | 39 | 139 | 62 | 0 | 5 | 0 | 0 | 0 | 771 | 73 | 487 |
| Aug* | 274 | 240 | 0 | 39 | 139 | 62 | 0 | 5 | 0 | 0 | 0 | 759 | 86 | 572 |
| Sep* | 263 | 239 | 0 | 39 | 139 | 62 | 0 | 5 | 0 | 0 | 0 | 748 | 98 | 651 |
| Oct* | 254 | 239 | 0 | 39 | 139 | 62 | 0 | 5 | 0 | 0 | 0 | 737 | 108 | 723 |
| Nov* | 0 | 482 | 0 | 239 | 0 | 0 | 0 | 5 | 238 | 0 | 0 | 727 | 91 | 609 |
| Dec* | 0 | 243 | 0 | 239 | 0 | 0 | 0 | 5 | 0 | 0 | 0 | 488 | 66 | 442 |
| AVERAGE | | | | | | | | | | | | 654 | 83 | 556 |

*Only applicable if a system of three matings in two years is practised*

## CONTRIBUTION OF ANIMAL CLASSES TO LSU'S

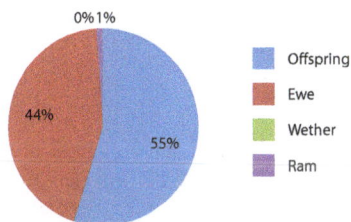

Stocking rates in terms of heads on the farm varied from a minimum 488 of to a maximum of 778. Ewes contributed 44% to the LSU values and the offspring 55% with 1% from rams. The gross margin for income was R491.58/SSU and R273.10/ha.

### 6.3.3.3 *Highveld Ile De France sheep*

The production system as described for Damara sheep, using the phenotypic and agro-economic values recorded for lle De France sheep at Schoongezicht on the Highveld, as tabulated in Table 6.13 was applied for the lle De France sheep. The income and gross margins are summarised in Table 6.16 for lle De France sheep at Schoongezicht.

Table 6.16 *Income and gross margins for Ile De France sheep on the Highveld*

|  | RAND | R/SSU | R/ha |
|---|---|---|---|
| **INCOME** | **237 250** | **427.05** | **237.25** |
| Animals | 198 933 | 358.08 | 198.93 |
| Products | 38 317 | 68.97 | 38.32 |
|  |  |  |  |
| **EXPENSES** | **28 803** | **51.85** | **28.80** |
| Animals | 369 | 0.66 | 0.37 |
| Feed | 0 | 0.00 | 0.00 |
| Shearing | 4 103 | 7.39 | 4.10 |
| Marketing | 13 220 | 23.80 | 13.22 |
| Health | 5 556 | 10.00 | 5.56 |
| Sundry | 5 556 | 10.00 | 5.56 |
|  |  |  |  |
| **GROSS MARGIN** | **208 447** | **375.20** | **208.45** |

**CLIP SUMMARY**

|  | Greasy wool (kg) | Clean wool (kg) | % |
|---|---|---|---|
| Lamb | 186 | 115 | 15 |
| Hogget | 316 | 212 | 28 |
| Main | 647 | 434 | 57 |
| **Total** | **1149** | **761** | **100** |

Stocking rates in terms of heads on the farm varied from a minimum of 460 to a maximum of 646. Ewes contributed 44% to the LSU values and the offspring 55% with 1% from rams. The gross margin for income was R375.20/SSU and R208.45/ha.

For ease of comparison, the income and gross margins for the three breeds at Schoongezicht on the Highveld are summarised in Table 6.17.

Table 6.17. *Summary of income and gross margins for the three breeds at Schoongezicht on the Highveld*

|  | Ile De France | Meatmaster | Damara |
|---|---|---|---|
| Income | 208 447 | 302 609 | 237 662 |
| Income/SSU | 427.05 | 544.70 | 427.79 |
| Income/ha | 237.25 | 302.61 | 237.66 |
|  |  |  |  |
| Gross margin | 208 447 | 273 101 | 211 437 |
| Gross margin/SSU | 375.20 | 491.58 | 380.59 |
| Gross margin/ha | 208.45 | 273.10 | 211.44 |

Figure 6.7 provides a graphical summary of the gross margin income values per SSU for the three breeds at Schoongezicht on the Highveld.

Fig 6.7. *Gross margin per SSU for the three breeds at Schoongezicht on the Highveld*

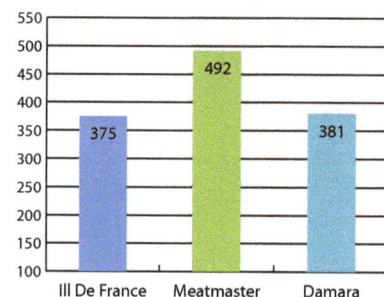

The relative comparative economic performance of Meatmaster and other sheep breeds at the three localities are summarised in Table 6.18.

*Table 6.18  Relative economic performance of breeds at the three localities (Meatmaster = 100)*

| Locality | Meatmaster gross margin | Damara | Dorper | SAMM | Ile De France | Afrino |
|---|---|---|---|---|---|---|
| Eastern Cape | 100.0% | 81.0% | 80.8% | 66.7% | - | 74.0% |
| Keetmanshoop | 100.0% | 67.0% | 71.0% | - | - | - |
| Highveld | 100.0% | 77.4% | - | - | 76.3% | - |
|  |  |  |  |  |  |  |
| **Average** | **100%** | **75.1%** | **75.9%** | **66.7%** | **76.3%** | **74.0%** |

Table 6.18 was compiled in an attempt to minimise the environmental effect on the comparisons between the vastly different agro-ecosystems involved were data of different breeds were recorded. The results confirm that different populations of the Meatmaster breed maintain an agro-economic edge of between 20% and 25% in gross margin compared to its parent and other breeds. Furthermore the Damara breed compared well to other breeds when a system of 3 matings in two years is practised. The lower value registered by the Damara at Keetmanshoop is due to the fact that the calculation based on a once a year mating in the case of Keetmanshoop deprived the Damara breed of the opportunity to register its ability to lamb at shorter inter lambing periods. The results in terms of the other breeds are in line with the initial findings of Snyman et al. (2000) that there were no significant differences in the long term gross income between a number of breeds which included the Dorper and the Afrino.

## 6.4  Conclusions

• The Meatmaster is an exceptional breed with very unique qualities. The results confirm that the Meatmaster breed does in fact have the production potential to outperform its parent breeds in extensive farming conditions.
• The Meatmaster breed maintained its superior economic performance with respect to its parent breeds at three localities of substantially different agro-ecosystems as well as in different management systems.
• Different populations of the Meatmaster breed performed substantially better than its parent breeds at all the localities.
• It can be reasonably expected that the Meatmaster sheep breed will gain a considerable market share worldwide, due to its agro-economic production potential, within a relative short period of time. The breed is already well

known in Southern Africa, Australia, Brazil and other parts of the world.
• The potential of the Meatmaster breed to perform in a 3 matings in two years production system should be tested by further research at localities conducive to such a production system. Based on the production parameters used at La Rochelle in the Eastern Cape for the input to the SM2000 program, the Meatmaster breed would yield a gross income margin of R720.21/SSU and R400.12/ha with stocking rates varying from a min of 419 heads to a max of 841. This is substantially above the calculated value of R521.80/SSU and R289.89/ha achieved by the current production system practised. It is also not necessary to maintain a replacement rate of 20% p.a. for Meatmaster sheep as the breed seems capable of maintaining an extended productive life. Eight-year-old ewes maintained acceptable production capacity on the Highveld.

## 6.5  Summative conclusions on the evaluation of the Meatmaster

Genetic and production characterization reveal the identity and tremendous potential of the locally developed composite Meatmaster sheep breed:

• The Meatmaster sheep breed was created and established in South Africa without state support and is a remarkable achievement as a result of the innovative approach by a number of South African sheep breeders.
• The establishment of the Meatmaster breed was highly successful in practical terms as the new Meatmaster sheep breed was formally recognized and proclaimed as a locally developed sheep breed in terms of the Animal Improvement Act by publication in the Government Gazette.

- Meatmaster breeders were effectively organized by the establishment of the Meatmaster Sheep Breeders' Society of South Africa which is already the 4th largest breeders' society amongst the ten mutton and dual purpose sheep breed societies in South Africa.
- Meatmaster breed standards are completely unique and are based on agro-economic considerations and functional efficiency which is conducive to future genetic improvement of the breed.
- There is a worldwide need for an easy-care, hair type, meat producing sheep. The Meatmaster is not a fat tailed type of sheep and provides an acceptable and marketable carcass to the consumer.
- The phenotypic description and linear body measurements of Meatmaster sheep recorded can serve as a reference point for further research.
- No previous research was done on Meatmaster sheep and there were no scientific publications available on Meatmaster sheep.
- The results from the genetic characterization of the Meatmaster breed based on micro-satellite analysis contributes to the global genetic database.
- The Meatmaster breed shows promising levels of genetic identity for a young breed and a high level of genetic diversity has been captured in the breed.
- The presence of high allelic diversity in the Meatmaster will reduce the likelihood of inbreeding depression in the short term and will be of benefit in providing diversity to use for specific applications such as parental studies.
- The genetic profile of the Meatmaster at its onset will provide a valuable reference point against which possible future genetic drift can be measured.
- The characterization of the production potential of the Meatmaster sheep provides valuable breed specific phenotypic values for a number of important production traits.
- The low phenotypic values established for the economically very important production traits of age at first lambing, short interlambing periods and low lamb mortality rates contributes substantially to the reproduction and production potential of the Meatmaster breed.
- Phenotypic values established for the production traits of weaning weight and post weaning weight of the Meatmaster breed provides

an important contribution to breed specific breeding values.
- Genetic trends evaluated for the Meatmaster breed between 1999 and 2009 confirm substantial genetic progress in the breed with respect to direct weaning weight, post weaning growth rate and revenue breeding values.
- The productive performance of the Meatmaster breed established at different locations in substantially different ecosystems confirms the adaptability of the new breed.
- The agro-economic evaluation of the Meatmaster breed confirms that the Meatmaster breed can outperform its parent breeds in extensive farming conditions.
- The Cape Glover quality skin of the Meatmaster sheep which present itself in innumerable colour patterns could become an additional source of income in the manufacturing of exotic sheep skin products.
- The Meatmaster breed registered superior economic performance with respect to its parent breeds at three localities of substantially different agro-ecosystems and in different management systems.
- The establishment of the Meatmaster breed based on the inclusion of the indigenous Damara breed as the main contributor to its genetic composition proved the importance of the conservation of well adapted indigenous breeds once again.

## 6.6    Recommendations for further research on the Meatmaster and indigenous sheep breeds

Due to the comprehensive nature of sheep breed establishment and the dearth of scientific information available on indigenous sheep breeds further research is recommended:

- Further research on the new Meatmaster sheep breed as well the indigenous Damara breed is highly recommended. No scientific publications are available in respect of the Damara breed with the inherent qualities to make a substantial contribution to the improvement of the production potential of the national and international small stock industry.
- The potential of the Meatmaster breed to perform in a 3 matings in two years production system should be investigated at localities conducive to such production systems.

- The true agro-economic value attributable to Meatmaster breed standards should be evaluated by further research.
- Meatmaster carcass characteristics and its potential to provide organically produced mutton to a growing world market.
- Feasibility of the development of new marketable skin products from the Cape Glover quality of Meatmaster and Damara skins available in South Africa.
- Life-time assessment of profitability in other diverse agro-ecosystems.
- Adaptability of the Meatmaster to agro-eco-systems in tropical environments.
- The potential of the Meatmaster to contribute to improved productivity of sheep farming on the African continent, Australasia and South America.
- Comparison of linear body measurements of Meatmaster sheep in different agro-ecosystems.
- The establishment of nucleus flocks by the National Department of Agriculture of Namaqua Afrikaner x Polled Wiltshire Horn and Damara x Romanov composites for future evaluation and research.
- Comprehensive phenotypic characterization of the Meatmaster breed and the establishment of Meatmaster breeding values.

# References

Act No.62 of 1998. *The Animal Improvement Act.* Government Gazette. Government Printer, Pretoria.

Arora R and Bhatia S. 2004. *Genetic structure of Muzzafarnagri sheep based on micro-satellite analysis.* Small Rumin. Res. 54: 227-230.

Alberts C. 2007. *Facts on the Meatmaster.* Go Farming Vol.2, No.1: 17-19.

Anderson CJ. 1856. *Lake Ngami: Exploration and discoveries during four years of wandering in the wilds of South West Africa.* Hurst and Blackett, London.

Baker CMA, Manwell C, Labrisky RF and Harper JA. 1966. *Molecular genetics of avian proteins – V. Egg, blood and tissue proteins of the Ringnecked pheasant, Phusianus colchicus.* L. Comp. Biochem. Phyiol. 17: 467-499.

Baker RL, Mwamachi DM, Audho JO, Aduda, EO and Thorpe W. 1999. *Genetic resistance to gastro-intestinal nematode parasites in Red Maasai, Dorper and Red Maasai x Dorper ewes in the sub-humid tropics.* Animal Science (UK) 69(2): 335-344.

Beaumont MA and Bruford MW. (1999). *Microsatellites in conservation genetics. In "Microsatellites: Evolution and Application"* (D.B. Goldstein & C. Schlotterer, Eds.), pp. 165-183, Oxford University Press, New York.

Blackmore DW, McGuillard LD and Lush JL. 1958. *Genetic relationship between body measurements at three ages in Holstein.* J. Dairy Sci. 41: 1045-1049.

Bonsma JC. 1980. *Livestock production: a global approach.* Cape Town, South Africa, Tafelberg Publishers.

Bonsma JC.1983. *Man must measure livestock production.* Cody. Wyoming. USA. Agri books.

Bourdon RM, 2000. *Understanding Animal breeding.* 2nd Ed. Prentice Hall. NJ.

Brown JE, Brown CJ and Butts WT. 1973. *Evaluating relationships among immature measures of size, shape and performance on beef bulls 1; principle component as measures of size and shape in young Hereford and Angus bulls.* J. Anim. Sci. 36: 1010-1020.

Buchanan FC, Swarbrick PA and Crawford AM. 1991. *Ovine dinucleotide repeats polymorphisms at the MAF65 locus.* Anim. Genet. 23: 85.

Buchanan FC and Crawford AM. 1992. *Ovine dinucleotide repeat polymorphism at the MAF214 locus.* Anim. Genet. 23: 394.

Buduram P. 2004. *Genetic characterization of Southern African sheep breeds using DNA markers.* MSc Dissertation. University of the Free State.

Burgess M. 2006. *Engineering Africa's ultimate veldt sheep.* Farmers Weekly 96035: 40-42.

Campbell QP. 1974. *A study of breeding problems in Dorper Sheep.* DSc Agric. Thesis. U.O.V.S.

Campbell QP. 1983. *Seleksie en teeltbeginsels by skape.* Vleisskape. B.1. Boerdery in Suid-Afrika. Departement van Landbou. Pretoria.

Campbell QP. 1986. *The Dorper - A success story of livestock improvement. Research Highlights.* Animal Production. Department of Agriculture and Water Supply. 27-28.

Campbell QP. 1988. *Eienskappe en faktore wat die funksionele doeltreffendheid van die Dorper kan beinvloed.* Dorpernuus. 14.

Campbell QP. 1995. *The indigenous sheep and goat breeds of South Africa.* Dreyer Printers and Publishers.

Cilliers J. 2000. *Potslagters sal teëspoed kry met ranteskape.* Landbouweekblad 1167: 44-46.

Cloete SWP, Snyman MA and Herselman MJ. 2000. *Productive performance of Dorper sheep.* Small Rumin. Res., 36, 119-135.

Cloete SWP, Cloete JJE, Herselman MJ and Hoffman LC. 2004. *Relative performance and efficiency of five Merino and Merino-type dam lines in a terminal crossbreeding system with Dormer or Suffolk sires.* S. Afr. J. Anim. Sci. 34, 135-143.

Collins J and Conington J. 2010. *Breeding easier-managed sheep.* SAC. Edinburgh. Scotland. www.sac.ac.uk/mainrep/pdfs/conningtoneasysheepreport.pdf (retrieved 21.09.2010).

Collett CR. 2003. *Balance the key to Meatmaster success.* Journal of the Damara and Meatmaster Society. Vol 7.

Collett CR. 2005. *Sustainable farming and successful livestock breeding.* (2005/08/04). http://www.collettgroup.com

Collett CR. 2007. *Meatmaster Breed Details.* (2007/02/20) http://clynton.collettgroup.com/meatmaster_sheep.html

Crawford AM, Dodds KG, Ede AJ, Montgomery GW, Garmonsway HG, Beattie AE, Davies K, Maddoz JF, Kappes SW, Stone RT, Nguyen TC, Penty JM, Lord EA, Broom JA and Buitkamp J. 1995. *An autosomal genetic linkage map of the sheep genome.* Genetics 140: 703-724.

DAGRIS. 2005. *Domestic Animal Genetic Resources Information System (DAGRIS).* (eds)

Ed Rege, Workneh Ayalew, Ephrem Getahun). ILRI (International Livestock Research Institute), Addis Ababa, Ethiopia. http://dagris.ilri.cgiar.org/dagris

Dickerson GE. 1978. *Animal size and efficiency: Basic concepts.* Anim. Prod. 27: 367-379.

Duguma G, Cloete SWP, Schoeman SJ and Jordaan GF. 2002. *Genetic parameters of testicular measurements in Merino rams and the influence of scrotal circumference on total flock fertility.* S. Afr. J. Anim. Sci. 32: 76-80.

Du Toit, CM. 2005. *Geelbeksdam Meatmaster Stud.* (2005/04/28). http://www.geelbeksdam.com/meatmaster.html

Du Toit CM and Du Toit HS. 2007. *Meatmaster Opleidingshandleiding.* Meatmaster Skaaptelersgenootskap van SA. Brandfort.

Du Toit D. 2007 Ed. *The Damara of Southern Africa.* Dawie du Toit, Prieska, South Africa.

El-Arian MN, Salama MA and Heba AM. 2008. *Estimation of genetic parameters and breeding values for growth traits on Romanov lambs in Egypt.* J.Agric.Sci. Mansoura Univ., 33(4), 2569-2576.

Erasmus LS, De Kock JA and Grobler JW 1983. *Slaglamproduksie in die Suid-Kaap.* Els. J. 7, 13-32. In Afrikaans.

Essien A and Adescope OM. 2003. *Linear body measurements of N'dama calves at 12 months in a South Western zone of Nigeria.* Livest. Res. Rural Devel. (15) 4.

Excoffier L, Laval G and Schneider S. 2005. Arlequin, version 3.0: *An Integrated Software Package for Population Genetics Data Analysis.* Computational and Molecular Population Genetics Laboratory (CMPG), Institute of Zoology, University of Berne, Switzerland.

Falush D, Stephens M and Pritchard JK. 2003. *Inference of population structure: extensions to linked loci and correlated allele frequencies.* Genetics 164: 1567-1587.

Fell HR. 1988. *Intensive Sheep Management* (2nd Ed). Farming

Press Books. Page Bros (Norwich) Ltd. UK.

Fourie AJ and Cloete SWP. 1993. *Reproductive performance of commercial Merino, Dohne Merino and SA Mutton Merino flocks in the Southern Cape*. S. Afr. J. Anim. Sci. 23, 104-110.

Fourie PJ, Neser FWC, Olivier JJ and van der Westhuizen C. 2002. *Relationship between production performance, visual appraisal and body measurements of young Dorper rams*. S. Afr. J. Anim. Sci. 32 (4): 256-262.

Fourie PJ, Schwalbach LM, Neser FWC and Greyling JPC. 2005. *Relationship between body measurements and serum testosterone levels of Dorper rams*. Small Rumin. Res. 56: 75-80.

Gilmour A, Thompson B, Cullis S, Welham SJ, 2000. *AS-REML Reference Manual*. Orange 2800. Australia: NSW Agriculture.

Goodman SJ. 1997. *RSTCALC: A collection of computer programs for calculating estimates of genetic differentiation from micro-satellite data and determining their significance*. Mol. Ecol. 6: 881-885.

Government Gazette. 2007. No. 29898. 25 May 2007. Government Printer, Pretoria. R450: 3-6.

Government Gazette. 2009. No 32601. 2 October 2009. Government Printer, Pretoria. R935:4-7.

Greef JC, Roux CZ, Wyma GA. 1990. *Lifetime meat production from six different F1 crossbred ewes*. S. Afr. J. Anim. Sci. 20, 71-77.

Hammond K. 2000. *A Global Strategy for the Development of Animal Breeding Programmes in Lower-Input Production Environments*. Animal Genetic Resources, Animal Production and Health Division, FAO, Rome, Italy.

Hartl DL and Clark AG. 1989. *Principles of Population Genetics*. Sinauer Associates, Sunderland, MA.

Haselholt M. 1969. *Serum protein polymorphisms in swine electrophoretic identification*. Genetics and Application. Munksgaard, Copenhagen, pp 48-52.

Herselman MJ, Sahlu T, Hart SP, and Goetsch AL. 1998. *Energy expenditure by dry and lactating Alpine does estimated by entry rate of carbon dioxide*. J. Dairy Sci. 81:2469-2474.

Herselman MJ. 2002. *SM2000-programme: A simulation model for the calculation of profitability of different small stock enterprises*. http://gadi.agric.za

Hofmeyr I. 2001. *Breeders are crossing Damaras with other mutton sheep in quest of the ultimate meat machine*. Farmers Weekly. 13 April 2001.

Hulme DJ, Silk JP, Redwin JM, Barendse W and Beh KJ. 1994. *Ten polymorphic ovine microsatellites*. Anim. Genet. 25: 434-435.

International Livestock Research Institute (ILRI) Website. 2006. *Capacity building for sustainable use of animal genetic resources in developing countries*. Animal Genetic Training Resources. (2006/08/15). http://www.ilri.org

Kemp LM, Brezinsky L and Teale AJ. 1993. *A panel of bovine, ovine and caprine polymorphic micro-satellites*. Anim. Gen. et 25: 363-365.

Kim KS, Yeo JS, Lee JW, Kim JW and Choi CB. 2002. *Genetic diversity of goats from Korea and China using micro-satellite analysis*. Asian-Aust. J. Anim. Sci. 15: 461-465.

Kossarek LM and Grosse WM. 1993. *Bovine dinucleotide repeat polymorphism*. J. Anim. Sci. 71: 3175.

Kotze A, Swart H, Grobler JP and Nemaangani A. 2004. *A genetic profile of the Kalahari Red goat breed from southern Africa*. S.Afr. J. Anim. Sci. 34: 10-12.

Kunene N, Nesamvuni EA, Fossey A. 2007. *Characterisation of Zulu (Nguni) sheep using linear body measurements and some environmental factors affecting these measurements*.

S. Afr. J. Anim. Sci. 37: 11-20.

Lasley JF. 1972. *Genetics of Livestock Improvement, 2nd Ed*. Prentice-Hall, Inc., Englewood Cliffs. N.J.

Lategan D. 2002. *Dorpers in die nuwe eeu*. Dorperskaaptelersgenootskap van SA.

Littell RC. Freud RJ. Spector PC. 1991. *SAS-system for linear models, 3rd Ed*., Cary, NC.

Luikart G, Biju-Duval MP, Ertugrul O, Zagdsuren Y, Maudet C and Taberlet P. (1999). *Power of 22 microsatellite markers in fluorescent multiplexes for parentage testing in goats (Capra hircus)*. Anim. Genet. 30: 431-438.

Maiwashe AN. 2000. *The value of recording body measurements in beef cattle*. MSc (Agric) dissertation, University of the Free State.

McManus C, Louvandini H and Paiva S. 2008. *Examples of different aspects of adaptive fitness, how they can be measured and possible proxi-indicators*. In: Report on the FAO/WAAP workshop on production environment descriptors for animal genetic resources report. Eds. Pilling D, Rischkowsky B and Scherf B, Caprarola,
Italy, 6 – 8 May, 2008.

Matika O, van Wyk JB, Erasmus GJ and Baker RL. 2001. *Phenotypic and genetic relationships between lamb and ewe traits for the Sabi sheep of Zimbabwe*. S. Afr. J. Anim. Sci. 2001, 31(3).

Matika O, van Wyk JB, Erasmus GJ and Baker RL. 2003. *A description of growth, carcass and reproductive traits of Sabi sheep in Zimbabwe*. Small Rumin. Res. 48 (2), 119-126.

Meatmaster Sheep Breeders' Society of SA. 2007. *Meatmaster Sheep Breeders*. (2007/01/16). http://eng.studbreeder.com/sheepbreeders_meatmaster.html

Michalakis Y and Excoffier L. 1996. *A genetic estimation of population subdivision using distances between alleles with special reference for micro-satellite loci*. Genetics 142: 1061-1064.

Miller SA, Dykes DD and Polesky HF. 1998. *A simple salting-out procedure for extracting DNA from human nucleated cells*. Nuc. Acids. Res. 16: 1215.

Muller Y. 2007. *Pioneering new sheep breed*. Meyerton Ster 8-12 Jan 2007: 5.

Nei M. 1972. *Genetic distance between populations*. The American Naturalist 106 (949): 283-292.

Nei M. 1987. *Molecular Evolutionary Genetics*. Columbia University Press, New York, NY, USA.

Oklahoma State University. 2007. *Hairsheep breeds*. (2007/05/10).http://www.okstate.edu/breeds/sheep/hair/htm

Norušis MJ. 2003. SPSS®. *Statistical Procedures Companion*. Prentice Hall. USR. NJ.

Olivier JJ. 2007. *Performance Testing of Small Stock*. ARC Livestock Business Division. Stellenbosch.

Olivier JJ, Cloete SWP, 2006. *Genetic Analysis of the South African Dorper Sheep*. Proc. 7th World Cong. Gen. Appl. Livest. Prod., Bello Horizonte, Brazil: Communication, pp. 04-10.

Olivier JJ., Cloete, SWP, 2007. *Genetic change in some South African Merino studs participating in the Merino plan*. Proc. Assoc. Advmnt Anim. Breed. Gen. 17, 324–327.

Ota T. 1993. *Dispan: genetic distance and phylogenetic analysis*. Pennsylvania State University, PA, USA.

Pallant J. 2007. *A step by step guide to data analysis using SPSS for windows*. 3rd Ed. McGraw Hill. New York. USA.

Park S. 2001. *Micro-satellite Toolkit*. (2005/04/28) http://oscar.gen.tcd.ie/sdepark/ms-toolkit

**95**

Peters FW. 2006. *Fecundity and productivity.* Meatmaster compared to Damara and Ile de France. Meatmaster Newsletter.

Peters FW. 2007. *Welcome to Meinfred Meatmasters.* (2007/01/16). http://meinfredmeatmaster.studbreeder.com

Peters FW. 2009. *Constitution of the Meatmaster Sheep Breeders' Society of South Africa.*

Peters FW. 2009. *By-laws to the Constitution of the Meatmaster Sheep Breeders' Society of South Africa.*

Peters FW. 2010. *Meatmaster sheep breed establishment and development based on agroeconomic principles, national breed standards, phenotypic and genetic characterization.* Meatmaster Journal Vol 1:17-18. Meatmaster Skaaptelersgenootskap van SA. Bloemfontein.

Peters FW, Kotze A, van der Bank FH, Soma P and Grobler JP. 2010. *Genetic profile of the locally developed Meatmaster sheep breed in South Africa based on microsatellite analysis.* Small Rumin. Res. 90: 101-108.

Philipsson, J., Rege, J.E.O. & Mwai, O. 2006. *Sustainable breeding programmes for tropical farming systems.* In: ILRI-SLU 2006. Animal Genetics Training Resource, CD Version. 2, eds J.M. Ojango, B. Malmfors & O. Mwai. ILRI (International Livestock Research Institute), Nairobi, Kenya, and SLU (Swedish University of Agricultural Sciences), Uppsala, Sweden.

Pritchard JK, Stephens M, Rosenberg NA and Donnelly P. 2000. *Association mapping in structured populations.* Am. J. Hum. Gen. 67: 170-181.

Quiroz J, Martinez AM, Zaragozac L, Perezgrovasc R, Vega-Plad JL and Delgado JV. 2008. *Genetic characterization of the autochthonous sheep populations from Chiapas, Mexico.* Livest. Sci. 116: 156-161.

Ramsey K, Harris L and Kotze A. 2000. *Landrace Breeds: South Africa's indigenous and locally developed farm animals.* Farm Animal Conservation Trust, Pretoria.

Rannala B and Mountain JL. 1997. *Detecting immigration by using multilocus genotypes.* Proc. Natl. Acad. Sci. USA. 94: 9197-9201.

Rendo F, Iriondo M, Jugo BM, Mazón LI, Aguirre A, Vicario A and Estonba A. 2004. *Tracking diversity and differentiation in six sheep breeds from the North Iberian Peninsula through DNA variation.* Small Rumin. Res. 52: 195–202.

Reynecke DP, van Wyk JA , Gummow B, Dorny P and Boomker J. 2009. (online). *Validation of the FAMACHA© eye colour chart using sensitivity/specificity analysis on two South African sheep farms.* Veterinary Parasitology (online 6 September 2009).

Rice WR. 1989. *Analyzing tables of statistical tests.* Evolution 43: 223–225.

Russel D. 2010. *Genelink Meatmaster sheep at Parilla in the Mallee of South Australia.* Meatmaster Journal Vol 1: 12-13. Meatmaster Skaaptelersgenootskap van SA. Bloemfontein.

SA Studbook. 2007. *Sheep breeders.* (2007/02/2 http://www.studbook.co.za/telers.php?ShowSoc

Saitbekova N, Schlapfer J, Dolf G and Gaillard C. 2001. *Genetic relationships in Swiss sheep breeds based on micro-satellite analysis.* J. Anim. Breed. Genet. 118: 379–387.

Sargent J. 2000. *Genetic Variation in Blood Proteins within and differentiation between 19 sheep breeds from Southern Africa.* MSc Dissertation. RAU.

SAS, 1996. *SAS Procedures Guide, (Version 6, 3rd ed.), SAS Institute.* Inc., Cary, N.C., USA.

Schoeman SJ and Combrink GC. 1987. *Testicular development in Dorper, Dohne Merino and crossbred rams.* S. Afr. J. Anim. Sci. 17: 22-26.

Schoeman SJ, De Wet R, Botha MA and van der Merwe CA. 1995. *Comparative assessment of biological efficiency of crossbred lambs from two composite lines and Dorper sheep.* Small Rumin. Res. 16, 61-67.

Schoeman SJ. 1996. *A preliminary assessment of predictive measures for the expression of weaning efficiency in sheep.* S. Afr. J. Anim. Sci. 26(2): 47-49.

Schoeman SJ, Cloete SWP, Olivier JJ, 2010. *Returns on investment in sheep and goat breeding in South Africa.* Livest. Sci. 130, 70-82.

Schoenian S. 2009. *A-Z of sheep breeds.* (2009/12/28). http://www.sheep101.info

Scholtz MM. 2005. *The role of research and a seed stock industry in the in situ conservation of livestock genetic resources.* Proceedings of 4th All Africa Conf. Anim. Agric. Eds. JEO. Rege, AM Nyamu and D Sendalo. Tanzania.

Scholtz MM, Furstenburg D, Maiwashe A, Makgahlela ML, Theron HE and van der Westhuizen J. 2010. *Environmental-genotype responses in livestock to global warming: A Southern African perspective.* S. Afr. J. Anim. Sci. (In press)

Scholtz MM, Spickett AM, Lombard PE and Enslin CB. 1991. *The effect of tick infestation on the productivity of cows of three breeds of cattle.* Onderstepoort J. Vet. Res., 58, 71 - 74.

Seymore DJ. 1937. *Hulpboek vir Boere in Suid Afrika.* Dept van Landbou en Bosbou. Staatsdrukker. Pretoria.

Slatkin M. 1995. *A measure of population sub-division based on micro-satellite allele frequencies.* Genetics 139: 457-462.

Sneath PHA and Sokal RR. 1973. *Numerical Taxonomy.* Freeman, San Francisco.

Snyman MA, Olivier JJ, Cloete JAN, 1993. *Productive and reproductive performance of Namaqua Afrikaner sheep.* Karoo Agric. 5 (2), 21-24.

Snyman MA, Olivier JJ, Erasmus GJ, Van Wyk JB, 1997. *Genetic parameter estimates for total weight of lamb weaned in Afrino and Merino sheep.* Livest. Prod. Sci. 48, 111–116.

Snyman MA, Erasmus GJ, Van Wyk JB, Olivier JJ, 1998. *Genetic and phenotypic correlations among production and reproduction traits in Afrino sheep.* S. Afr. J. Anim. Sci. 28, 74-81.

Snyman MA, Herselman MJ, Cloete JAN and King BR.1999. *Bruto inkomste van Afrino- Dorper- en Merinoskape.* Grootfontein LOI. Middelburg. (2008/09/29) http://gadi.agric.za/articles/Snyman_MA/rasse.htm

Snyman MA, Herselman MJ, Cloete JAN and King BR. 2000. *Bruto marge van Afrino,Dorper en Merinoskape.* Grootfontein Agric 2 (2), 9-11.

Snyman MA, Olivier WJ, 2002. *Productive performance of hair and wool type Dorper sheep under extensive conditions.* Small Rumin. Res. 45, 17-23.

Snyman MA and Herselman MJ , 2005. *Comparison of productive and reproductive efficiency of Afrino, Dorper and Merino sheep in the False Upper Karoo.* S. Afr. J. Anim. Sci. 35 (2), 98-108.

Spencer CC, Neigel JE and Leberg PL. 2000. *Experimental evaluation of the usefulness of micro-satellite DNA for detecting demographic bottlenecks.* Mol. Ecol. 9: 1517–1528.

Spickett AM, de Klerk D, Enslin CB and Scholtz MM. 1989. *Resistance of Nguni, Bonsmara and Hereford cattle to ticks in a bushveld region of South Africa.* Onderstepoort J. Vet. Res. 56, 245 251.

Steffen P and Eggen A. 1993. *Isolation and mapping of polymorphic micro-satellites in cattle.* Anim. Genet. 24: 121-124.

Steyn JJ. 2006. *Meatmaster Teelprogram: Geleenthede om be-*

*trokke te raak.* Meatmaster Newsletter.

Tautz, D. and Renz, M., 1984. *Simple sequences are ubiquitous repetitive components of eukaryotic genomes.* Nucl. Acids Res. 12, 4127 –4138.

Taye M, Abebe G, Gizaw S, Lemma S, Mekoya A and Tibbo M, 2010: *Traditional management systems and linear body measurements of Washera sheep in the western highlands of the Amhara National Regional State, Ethiopia.* Livestock Research for Rural Development. 22(9),169. Retrieved October 17, 2010, from http://www.lrrd.org/lrrd22/9/taye22169.htm

Tibbo M. 2006. *Productivity and Health of Indigenous Sheep Breeds and Crossbreds in the Central Ethiopian Highlands.* PhD thesis. Swedish University of Agricultural Sciences. Uppsala.

Tomasco I, Wlasiuk G and Lessa EP. 2002. *Evaluation of polymorphism in ten micro-satellite loci in Uruguayan sheep flocks.* Genet. Mol. Biol. 25: 37–41.

Vaiman D, Mercier D, Moazami-Goudarzi K, Eggen A, Ciampolini RK, Lepingle A, Velma R, Kaukinnen J, Varvio SL, Martin P, Leveziel H and Guerin G. 1994. *A set of 99 cattle micro-satellites: characterization, synteny mapping and polymorphism.* Mamm. Genome 5: 288-297.

Van Marle-Köster E and Nel LH. 2003. *Genetic markers and*

*their application in livestock breeding in South Africa: A review.* S. Afr. J. Anim. Sci. 33, 1-10.

Van Rooyen C. 2009. *Rustenburg-skou: Meatmaster Nasionale rasverteenwoordigers.* Landbouweekblad. 13 Februarie 2009. 37.

Van Wyk JB, Fair MD, Cloete SWP, 2009. *Case study: the effect of inbreeding on the production and reproduction traits in the Elsenburg Dormer sheep stud.* Livest. Sci. 120, 218-224.

Van Wyk JB, Swanepoel JW, Cloete SWP, Olivier JJ, Delport GJ, 2008. *Across flock genetic parameter estimation for yearling body weight and fleece traits in the South African Dohne Merino population.* S. Afr. J. Anim. Sci. 38, 31–37.

Walker CH. 2010. *Easier managed sheep and beef cattle; simplified, profitable and productive sheep and beef farming.* A Nuffield Farming Scholarships Trust Report. www.nuffieldinternational.org/rep_pdf/1253801058C_H_Walker_Nuffield_Report_read_only.pdf (retrieved 22.09.2010)

Wollny CBA. 2003. *The need to conserve farm animal genetic resources in Africa. Should policy makers be concerned?* Ecological Ecomics, 45 (3): 341-351.

Wright S. 1965. *The interpretation of population structure by F-statistics with special regard to systems of mating.* Evoltion 19: 395-420.

# Appendix A   Membership list and contact details of Meatmaster sheep breeders in South Africa

**MEATMASTER SHEEP BREEDERS' SOCIETY OF SOUTH AFRICA**
**PO BOX 1060 BLOEMFONTEIN, 9300 - TEL: 051 410 0955 FAX: 051 448 4220**

## MEMBERSHIP LIST

| KKM | NAME | ADDRESS | TELEPHONE NO |
|---|---|---|---|
| SPB | ALASIL FARM<br>Tony Spolidoro<br>92 DERBY ROAD<br>KENSINGTON<br>2094 | PO BOX 1151<br>BEDFORDVIEW<br>2008 | 011 615 4841<br>011 615 5671 (F)<br>083 379 2406<br>gbarrable@mweb.co.za |
| | ANYSBERG BOERDERY<br>Johan Saaiman<br>Pierre Rousseau | PO BOX 418<br>LADISMITH<br>6655 | 028 551 1154 (T & F)<br>083 653 7310<br>anysbergbs@mweb.co.za |
| | BALIE JR | PO BOX 236<br>STEINKOPF<br>8244 | 027 721 8910<br>083 989 8910 |
| THB | BISSETT TH<br>Tommy<br>TOMBI MEATMASTERS | PO BOX 350<br>HENDRINA<br>1095 | 082 823 1262 |
| | BOONZAAIER C<br>INTERAGRI | POSTNET SUITE 68<br>PRIVATE BAG X2449<br>MOKOPANE<br>0600 | 082 806 5294<br>carl@interagri.co.za |
| | BREDENKAMP K<br>Kobus<br>DJR MEATMASTERS | PO BOX 341<br>HERTZOGVILLE<br>9482 | 082 897 0303<br>086 592 1770 (F)<br>kobusbredenkamp@gmail.com |
| CRC | COLLETT CR<br>Clynton<br>COLETT FARMING<br>SUPERIOR GENETICS | La Rochelle<br>PO BOX 122<br>BETHULIE,<br>9992 | 051 654 0538 (T)<br>086 643 0512 (F)<br>clynton@collettgroup.com<br>082 463 5936 |
| GJC | COETZER GJ<br>Gerrie<br>GREAT KEI MEATMASTERS | PO BOX 84<br>KEI MOUTH<br>5260 | 043 841 1139 (T & F)<br>083 654 4656<br>gerrie@keimouth.co.za |
| JJC | COMBRINCK MEATMASTERS<br>Johan | PO BOX 98<br>ORANIA<br>8752 | 053 207 0043 (T & F)<br>083 296 1166<br>johancombrinck@rocketmail.com |

| | | | |
|---|---|---|---|
| WJ | DE LANGE P<br>Piet<br>LELIEKRANS MEATMASTERS | PO BOX 84<br>CALVINIA<br>8190 | 027 341 2035 (T&F)<br>072 738 8621<br>pietneliadelange@gmail.com |
| | DERCKSEN H<br>Hennie | PO BOX 344<br>HERTZOGVILLE<br>9482 | 053 421 9363<br>hmdgarisma@intekom.co.za |
| AGD | DIEDERICKS AG<br>ORA MEATMASTER STUD | 1 O'REILLIEG STREET<br>HOPETOWN<br>8750 | 053 203 0119<br>082 332 6909<br>agd@webmail.co.za |
| DVB | DE VILLIERS JD<br>Jan<br>SIGNALBERG | PO BOX 1<br>GRUNAU<br>NAMIBIA | 00264 6326 2067 (T & F)<br>00264 8131 54745 (C)<br>jandev@mtcmobile.com.na |
| | DU PLESSIS JP<br>Jean<br>ELANDSNEK MEATMASTERS | PO BOX 134<br>HOPETOWN<br>8750 | 053 683 0013<br>082 064 5879<br>elandsnek@lantic.net |
| CC | DU PREEZ M<br>Mario | La Rochelle<br>PO BOX 122<br>BETHULIE<br>9992 | 051 654 0538 (T)<br>086 643 0512 (F)<br>083 458 8305<br>mario@collettgroup.com |
| RAE | ENGELBRECHT AR<br>HARRISDALE MEATMASTERS | PO BOX 89<br>UPINGTON<br>8800 | 083 259 1303<br>wesgor35@yahoo.com |
| JR | FOUCHE R<br>OLIVE BRANCH MEATMASTERS | PO BOX 63<br>GERDAU<br>2729 | 014492 ask 3803<br>018 673 0703 (F)<br>084 240 5269<br>jjfouche@msn.com |
| LPJ | FOURIE LPJ | PO BOX 166<br>LUCKHOFF<br>9982 | 082 313 4746<br>info@reedbuckridge.co.za |
| | GIAN BOERDERY BK<br>Gideon van der Westhuizen | PO BOX 184<br>KAKAMAS<br>8870 | 054 431 0458/9 (W)<br>027 662 1219 (Farm)<br>054 431 0454 (F)<br>082 827 8775<br>gideon@gvdw.biz |
| GWN | GLENWOOD MEATMASTERS<br>TNJ & HJL van der Walt<br>Tjaard & Henri<br>Glenwood | PO BOX 245<br>NELSPRUIT<br>1200 | 013 745 8105 (T & F)<br>082 440 7511<br>083 634 8877 |
| | JERUED MEATMASTERS<br>BOTHA K<br>Koetoe | PO BOX 898<br>KRUGERSDORP<br>1740 | 011 858 4000<br>072 112 8732<br>086 600 8434 (F)<br>koetoeb@actionford.co.za |

|  |  |  |  |
|---|---|---|---|
|  | JH GROBLER BOERDERY<br>Jan | PO BOX 216<br>ORKNEY<br>2620 | 018 441 1124<br>018 441 1126 (F)<br>083 654 2121<br>jangrobler@btbits.co.za |
| AA | HIGGS AS | PO BOX 38<br>HOPETOWN<br>8750 | 053 203 8018<br>083 744 2253<br>higgsas@webmail.co.za |
| BM | HIGGS W<br>Wim<br>BOKPUT MEATMASTERS | PO BOX 304<br>HOPETOWN<br>8750 | 053 203 0160 (F)<br>053 203 0352<br>082 775 5966<br>whiggshiqhpt@telkomsa.net |
|  | HUYSER DJB<br>Danie<br>DANKO | PO BOX 7025<br>MIEDERPARK<br>2527 | 018 291 1020<br>083 709 4117<br>danie.huyser@nwu.ac.za |
| PSJ | JANSE VAN RENSBURG PS<br>Pieter | PO BOX 2526<br>MONTANAPARK<br>0159 | 012 548 6815<br>012 548 0997 (F)<br>082 496 0622<br>dblt195@mweb.co.za |
|  | KAROO BOERDERY TRUST<br>Johan Moolman | PO BOX 44<br>BEAUFORT WES<br>6970 | 017 735 2546<br>082 899 3656<br>jzmoolman@xsinet.co.za |
|  | KIRSTEN S<br>Stephan | PO BOX 623<br>GRAAFF-REINET<br>9280 | 082 376 3270<br>stephan@ngunicattle.com |
| LJK | KLEYNHANS LJ<br>Louis<br>LOUCA | PO BOX 901<br>HEILBRON<br>9650 | 058 852 2576<br>086 539 9575<br>082 770 1112<br>louisjk@vodamail.co.za |
| DMM | KLEYNHANS LP<br>Flip<br>DUINEKROON MEATMASTERS | PO BOX 85<br>STILBAAI<br>6674 | 028 754 1643<br>082 821 3539<br>epkleinhans@telkomsa.net |
| DSB | KOTZE DA<br>Daniel<br>DANSAR | PO BOX 215<br>VANRHYNSDORP<br>8170 | 083 628 8989 |
| EX | KRUGER DW<br>Derick | PO BOX 7<br>MAASSTROOM<br>0623 | 076 034 8580<br>086 601 9871 (F)<br>derickk@mweb.co.za |
| JKI | KRUGER MI<br>Irene<br>JKI | PO BOX 6262<br>GREENHILLS<br>1767 | 011 412 3856/7<br>011 412 4715 (F)<br>082 902 3330<br>irenedomaine@icon.co.za |

|       | KEARNEY HA<br>OSVLEY BOERDERY | PO BOX 101<br>LOERIESFONTEIN<br>8185 | 027 662 1155 (T)<br>083 583 6168<br>irmari@hantam.co.za |
|-------|-------------------------------|--------------------------------------|---------------------------------------------------------|
|       | KEARNEY JF<br>Henk<br>NATURE'S MEATMASTERS | PO BOX 101<br>LOERIESFONTEIN<br>8185 | 027 662 1155 (T)<br>083 583 6168 |
| NJG   | KERN MEATMASTERS<br>Grobler NJ | PO BOX 223<br>DOUGLAS<br>8730 | 084 790 8191 |
|       | KOEKEMOER B<br>Burger | PO BOX 33<br>SKUINSDRIF<br>2851 | 083 230 5815<br>086 649 9023 (F)<br>ftdc@mweb.co.za |
| HE    | KOEKEMOER HJ<br>Hennie<br>DRINKWATER | PO BOX 60<br>HOPETOWN<br>8750 | 053 203 8232<br>082 468 6932<br>drinkwater@interexcel.co.za |
|       | LIEBENBERG CR<br>Riaan | PO BOX 624<br>UPINGTON<br>8800 | 083 383 7791 (Riaan)<br>082 372 6681 (Hannelie)<br>086 593 0190 (F)<br>crhliebenberg@vodamail.co.za |
|       | LK MEATMASTERS<br>GROUP OF 5 BREEDERS | PO BOX 162<br>LOERIESFONTEIN<br>8185 | 027 662 1155 (T)<br>083 583 6168 |
|       | LOUW N<br>Nico | PO BOX 184<br>LOERIESFONTEIN<br>8185 | 02762 ask 1240<br>082 774 7315 |
|       | MARITZ AWA<br>Riaan<br>PRAIRIE MEATMASTERS | PO BOX 343<br>POSTMASBURG<br>8420 | 05962 ask 1403<br>083 523 2123<br>ewalt.maritz@yahoo.com |
| JW    | MOHIMBA MEATMASTERS<br>MORRISON JW<br>Johnny | PO BOX 17317<br>BAINSVLEI<br>9338 | 051 445 2010<br>086 622 5158 (F)<br>083 383 2737<br>mohimba@telkomsa.net |
| HLJM  | MURRAY HLJ<br>Johan<br>PIENAARSPANSTOETERY | PO BOX 78<br>STRYDENBURG<br>8765 | 053 683 7003<br>082 801 4666 |
|       | NEL J<br>Johan<br>DRIEKOP MEATMASTERS | PO BOX 117<br>LOERIESFONTEIN<br>8185 | 027 6621 1505 |

| MF | PETERS FW<br>Freddie<br>MEINFRED MEATMASTERS | PO BOX 18<br>RANDVAAL<br>1873 | 016 365 5350<br>011 559 6161 (W)<br>016 365 5350 (F)<br>083 521 3990<br>freddiepeters@gmail.com |
|---|---|---|---|
| | PIETERSE SJ<br>Sarie<br>SARWIPI MEATMASTERS | PO BOX 475<br>GREYLINGSTAD<br>2415 | 082 903 4336<br>sarwipi@telkomsa.net |
| MOOI | PRETORIUS J<br>Johan<br>MOOIUITSIG | 11 PAPEGAAI STREET<br>RANT EN DAL<br>KRUGERSDORP 1751 | 011 660 5642<br>011 660 8868 (F)<br>082 496 2695<br>vaalkop@telkomsa.net |
| | ROSSOUW WA<br>Willem | PO BOX 18<br>PHILLIPOLIS<br>9970 | 051 773 7030 |
| DS | STEENKAMP DWM<br>Dennis<br>PARDEK | PO BOX 33<br>WILLISTON<br>8920 | 053 391 3059<br>053 391 3909 (F)<br>053 391 1903 (H)<br>082 336 7652<br>jos@kingsley.co.za |
| | SPANGENBERG PAL<br>Piet<br>VOSFONTEIN MEATMASTERS | PO BOX 151<br>LOERIESFONTEIN<br>8185 | 027 662 1202 (T)<br>083 2549 9801 |
| FWS | SPANGENBERG BOERDERY<br>EDM BPK (FWC & F)<br>Otto- OTRICKA | PO BOX 428<br>KEIMOES<br>8860 | 082 825 2715<br>054 461 2001 (T)<br>fspangenberg@telkomsa.net |
| | STEENKAMP JM<br>Jo | PO BOX 33<br>WILLISTON<br>8920 | 053 391 3059<br>053 391 3909 (F)<br>082 802 2389 |
| | STEENKAMP W | PO BOX 77<br>CALVINIA<br>8190 | 027 341 2549<br>027 341 1810<br>083 265 4594<br>wilgerboom@hantam.co.za |
| KI | STEYN JJ<br>Johan<br>BEN-TSON MEATMASTERS STUD | PO BOX 100600<br>RENOSTERSPRUIT<br>BLOEMFONTEIN<br>9326 | 051 441 7914 (W)<br>082 415 3999 – Johan Steyn<br>082 825 5091 – Fanie Steyn<br>083 564 1763 – JW Swanepoel<br>083 383 2739 – Johnny Morrison<br>ramsem@intekom.co.za<br>anipharm@telkomsa.net |
| JAS | SWIEGERS JA<br>Johannes | PO BOX 144<br>HOPETOWN<br>8750 | 053 203 8110<br>083 233 7848<br>jaswiegers@gmail.com |

| TY | TAYLOR ID<br>Ian<br>MIDAS MEATMASTERS | PO BOX 338<br>BEAUFORT WEST<br>6970 | 023 414 3397<br>086 634 1843 (F)<br>082 552 3221<br>istaylor@mweb.co.za |
|----|----|----|----|
| MM | VAN DER MERWE A (DR)<br>Andre<br>MILKY WAY MEATMASTERS | 8 BOSICA STREET<br>LOEVENSTEIN<br>7530 | 021 913 7190 (T & F)<br>083 556 2828<br>arvdm@sun.ac.za |
| | VAN DER MERWE M<br>Marcel<br>GLEN DOONE | PO BOX 27<br>LADY GREY 9755 | 051 603 7035<br>082 445 9277<br>lecram@lantic.net |
| | VAN DER POEL ANDRE<br>Andre<br>HOLANDIA MEATMASTERS | ZONKOLOL EXT 1973<br>CULLINAN<br>1000 | 012 736 0034<br>012 736 0064 (F) |
| DVDV | VAN DER VYVER D<br>Deon<br>MIERENFONTEIN | PO BOX 780<br>DURBANVILLE<br>7551 | 021 976 8619 (T&F)<br>082 551 4783<br>dlvdvyver@mweb.co.za |
| RN | VAN DER VYVER EG<br>FAMILIE TRUST<br>RANETTE STUD<br>Hannelie Pretorius | PO BOX 7356<br>FLAMWOOD<br>2572<br>PO BOX 215<br>VENTERSDORP<br>2710 | 018 462 5242<br>083 414 4840<br>083 236 8520 (Hannelie)<br>henlie@yebo.co.za |
| | VAN DER VYVER G<br>Spyker<br>HONORARY MEMBER | PO BOX 96<br>KROONDAL<br>0350 | 014 592 0346 (F)<br>083 372 9889<br>014 533 0508 |
| | VAN DER WESTHUIZEN AS<br>Kokkie | PO BOX 98<br>LOERIESFONTEIN<br>8185 | 027 662 1219 (T & F)<br>072 306 9192 |
| CVDW | VAN DER WESTHUIZEN CM<br>Chrismari<br>JOBSKRAAL | PO BOX 30<br>LOERIESFONTEIN<br>8185 | 02762 2340<br>xmaryvander@yahoo.co.uk |
| JVDW | VAN DER WESTHUIZEN J<br>JVDW MEATMASTERS | PO BOX 30<br>LOERIESFONTEIN<br>8185 | 027 622 313 or 3303<br>082 801 1895 |
| JVD | VAN DEVENTER JF<br>Jannie<br>BIANDI MEATMASTERS | PO BOX 2531<br>KLERKSDORP<br>2570 | 018 484 2512<br>018 484 6406 (F)<br>084 510 6203<br>072 683 9262 (Sanet)<br>jannie@biandi-simbras.co.za |
| EV | VERMEULEN ER<br>Edward<br>601388 | PO BOX 401<br>HOPETOWN<br>8750 | 086 661 0691 (F)<br>072 686 2462<br>edwardv@satb.co.za |

|     |                                                          |                                          |                                                                  |
|-----|----------------------------------------------------------|------------------------------------------|------------------------------------------------------------------|
|     | VERMEULEN SO<br>Okkie                                    | PO BOX 28<br>HOPETOWN<br>8750            | 072 686 2432<br>okkiev@vodamail.co.za<br>okkiev@pkfbfn.co.za     |
| PV  | VERMEULEN PvdW<br>Waltie<br>MARTINSPAN                   | PO BOX 170<br>HOPETOWN<br>8750           | 082 829 5395<br>waltie@cybertrade.co.za<br>waltie@mylink.co.za   |
| WV  | VERMEULEN W<br>Werner<br>TSAMA MEATMASTERS               | PO BOX 782<br>DIBENG<br>8463             | 078 575 1144<br>wv8343@gmail.com                                 |
|     | VISAGIE<br>Pine<br>DAGBREEK MEATMASTERS                  | PO BOX 4<br>LOERIESFONTEIN<br>8185       | 027 662 1355 (T)<br>027 662 1004                                 |
| DVS | VISSER DH<br>Danie<br>DIE BULT                           | PO BOX 42<br>KENHARDT<br>8900            | 05472 (Putsonderwater) 1712<br>083 390 0860<br>dhvisser@telkomsa.net |
|     | VLOK D<br>Deon<br>ROOIDAM MEATMASTERS                    | PO BOX 76<br>CALVINIA<br>8190            | 027 341 2597<br>072 474 7118<br>jdvlok@hantam.co.za              |
| MWM | VORSTER WLR<br>Willem<br>MELTON WOLD BOERDERY            | PO BOX 1<br>MELTONWOLD<br>7071           | 053 621 0906 (T&F)<br>082 442 9950<br>meltonwold@lantic.net      |
| K   | WESSELS KA<br>Karin<br>GRASVELDER MEATMASTERS            | PO BOX 272<br>FRANKFORT<br>9830          | 082 854 1491<br>kwessels@vodamail.co.za                          |
|     | WIID R<br>Roe                                            | PO BOX 212<br>HOPETOWN<br>8750           | 053 203 0167<br>082 787 9370                                     |
| ZFT | ZEEMAN CJ                                                | PO BOX 198<br>HOPETOWN<br>8750           | 053 203 8128<br>076 481 9633                                     |
|     | ZWIEGERS JAS                                             | PO BOX 124<br>HOPETOWN<br>8750           | 053 203 8148<br>082 783 6103<br>jaszwiegers@gmail.com            |

# Appendix B   Official proclamation of the Meatmaster sheep breed

## Appendix B1   Correspondence - Application for registration of the Meatmaster Breeders' Society and the Meatmaster sheep breed

**agriculture**

Department:
Agriculture
REPUBLIC OF SOUTH AFRICA

**Directorate: Animal and Aquaculture Production**
Private Bag X138, Pretoria, 0001
Delpen Building, % Annie Botha & Union Street, Riviera, 0084

**From:** Registrar: Animal Improvement
Mr Joel Mamabolo
**Tel:** +27 12 319 7424 **Fax:** +27 12 319 7570 / 7425
**E-mail:** JoelM@nda.agric.za **Website:** www.nda.agric.za
**Ref:**

15 May, 2006   -

The Meatmaster Sheep Breeders'
Of South Africa
P O Box 328
BRANDFORT
9400

**APPLICATION FOR REGISTRATION AS A BREED SOCIETY FOR MEATMASTERS**

Your application dated 28 April 2006 refers.

I hereby acknowledge receipt of the above-mentioned with thanks.  Please allow my office to review the application and will revert back to you as practical as possible.

Kind regards,

MR M J MAMABOLO
**REGISTRAR: ANIMAL IMPROVEMENT**

## Appendix B2  Proclamation of the Meatmaster sheep breed as a developing breed – Government Gazette No. 29898 of 25 May 2007

STAATSKOERANT. 25 MEI 2007                                        No. 29898   3

## GOVERNMENT NOTICES
## GOEWERMENTSKENNISGEWINGS

### DEPARTMENT OF AGRICULTURE
### DEPARTEMENT VAN LANDBOU

**No. R. 450**                                                              **25 May 2007**

### ANIMAL IMPROVEMENT ACT, 1998 (ACT No. 62 OF 1998)

### REGULATIONS: AMENDMENT

The Minister of Agriculture has, under section 2 of the Animal Improvement Act, 1998 (Act No.62 of 1998), made the regulation in the Schedule.

### SCHEDULE

#### Definations

1. In this Schedule "the Regulations" mean the Regulations published by Government Notice No.R.1682 of 21 November 2003, No.R.579 of 17 June 2005 and No.R.275 of 31 March 2006

#### Amendment of table 7 of the Regulations

2. The table in the annexure is hereby amended,

**Lulama Xingwana**
**Minister of Agriculture.**

TABLE 7 (d) - TABEL 7 ()

**BREEDS OF ANIMALS / RASSE VAN DIERE**
Breeds currently being evaluated

| Cattle / Beeste – Beef | Cattle/Beeste – Dairy | Goats / Bokke | Horses / Perde | Sheep /Skape |
|---|---|---|---|---|
| 1 | | 2 | 3 | 4 |
| Ankole<br>Afrigus<br>Mashona<br>S.A.Braford<br>Sanganer<br>Simbra | Swedish red | | S.A.Sport Horse<br>S.A.Miniature Horse | Meat Master Sheep<br>S.A.Milk Sheep |

## Appendix B3  Proclamation of the Meatmaster sheep breed as a locally adapted and regularly introduced breed – Government Gazette No. 32601 of 2 October 2009

**4**   No. 32601            GOVERNMENT GAZETTE, 2 OCTOBER 2009

## GOVERNMENT NOTICES
## GOEWERMENTSKENNISGEWINGS

### DEPARTMENT OF AGRICULTURE, FORESTRY AND FISHERIES
### DEPARTEMENT VAN LANDBOU, BOSBOU EN VISSERYE

No. R. 935                                                2 October 2009

### ANIMAL IMPROVEMENT ACT, 1998
### (ACT NO. 62 OF 1998)

### REGULATIONS: AMENDMENT

The Minister of Agriculture, Forestry and Fisheries, acting under section 2 of the Animal Improvement Act, 1998 (Act No. 62 of 1998), has made the following regulations in the Schedule.

### SCHEDULE
**Definition**

1.    In this Schedule "the Regulations means the Regulations published by Government Notices No. R. 1682 of 21 November 2003, as amended by Government Notices Nos. R. 579 of 17 June 2005 and R 275 of 31 March 2006.

**Substitution of Table 7 of the Regulations**

2.    The following table is hereby substituted for Table 1 of the Regulations.

6    No. 32601                                    GOVERNMENT GAZETTE, 2 OCTOBER 2009

## TABLE 7 (b) – TABEL 7 (b)

## BREEDS OF ANIMALS / RASSE VAN DIERE

2.    **Locally adapted and regularly introduced breeds**
      **(Other declared breeds / Ander verklaarde rasse)**

| Cattle / Beeste Beef / Vleis | Cattle / Beeste Dairy / Melk | Goats / Bokke | Horses / Perde | Sheep / Skape | Pigs / Varke |
|---|---|---|---|---|---|
| Angus | Ayrshire | Angora | American Quarter Horse | Border Leicester | Chester White |
| Boran | Dairy Shorthorn / Suiwel Shorthorn | British Alpine | Appaloosa | Corriedale | Duroc |
| Brahman | Jersey | Gorno Altai | Arab Horse / Arabierperd | Dorset Horn | Large Black / Groot Swart |
| Brangus | SA Dairy Swiss / SA Suiwel Switser | Saanen | Clydesdale | Hampshire | Large White / Groot Wit |
| Braunvieh | SA Friesland / SA Fries | Toggenberger | Connemara Pony / Connemara-ponie | Ile de France | Hampshire |
| Charolais | SA Guernsey | | English Halbblut | Meatmaster Sheep | Pietrain |
| Dexter | SA Holstein | | Friesian Horse / Friesperd | Merinolandsheep | Hamline |
| Galloway | | | Lipizzaner | Romanov | Robuster |
| Gelbvieh | | | Morgan horse / Morganperd | Southdown | SA Landrace / SA Landras |
| Limousin | | | Paint Horse | Suffolk | Welsh / Walliese |
| Pinzguaer | | | Percheron | | |
| Red Poll / Rooipoenskop | | | SA Hackney | | |
| Romagnola | | | SA Hackney pony / Hackney-ponie | | |
| SA Beefmaster | | | SA Warm Blood / SA Warmbloed | | |
| SA Braford | | | Saddler / Saalperd | | |
| SA Hereford | | | Shire | | |
| Santa Gertudis | | | Thoroughbred / Volboed | | |
| Senepol | | | Welsh Pony / Walliese ponie | | |
| Shorthorn | | | | | |
| Simbra | | | | | |
| Simmentaler | | | | | |
| South Devon | | | | | |
| Sussex | | | | | |
| Tuli | | | | | |
| Wagyu | | | | | |

| Ostrichies / Volstruis | Dogs | Others animals |
|---|---|---|
| | Anatolian Shepherd | SA Alpaca |
| | Boxer | |
| | SA Greyhound | |
| | SA Jack Russel | |

# Appendix C   NAMPO show - Meatmaster research poster - May 2007

## *Meatmaster* Sheep Breeding

FW Peters[1], FH van der Bank[1], A Kotze[2,3] and P Soma[4]

National Zoological Gardens of South Africa

[1] Department of Zoology, University of Johannesburg, PO Box 524, Auckland Park, Johannesburg 2006.
[2] National Zoological Gardens of South Africa, PO Box 754, Pretoria 0001.
[3] Department of Genetics, University of the Free State, PO Box 339, Bloemfontein 9300.
[4] Livestock Business Division, Agricultural Research Council, Private Bag X2, Irene 0062.

Durability and Fertility

**+**

Conformation

**=**

### VISION

A New Composite Landrace Breed

*Meatmaster*

| | | |
|---|---|---|
| Breed a population of Meatmaster Sheep which can provide breeding material to Meatmaster breeders and commercial farmers | Provide historic, genetic, phenotypic and agro-economic reference material on the Meatmaster Sheep Breed | Provide guidelines on breeding programs in order to achieve Meatmaster Breed Standards and optimum productivity |

## HISTORY

CR COLLETT   Dr W VERMEULEN   CM DU TOIT   FW PETERS   Dr J STEYN
Damara+Dorper   Damara+Van Rooy+Dorper   Damara+SA Mutton Merino   Damara+Ile De France   Damara+Wiltshire Horn

**PLUS OTHER INTERESTED BREEDERS**

INITIAL BREED NAMES:
DAMARA CROSSBRED SHEEP
ILE DA MARA
BOSRANDLER
MEATMASTER

ORGANISATION:   MEATMASTER COMMITTEE
DAMARA & MEATMASTER BREEDERS' SOCIETY

### PRESENT

- Meatmaster Sheep Breeders' Society of South Africa
- *Meatmaster* breed standards
- 20 Registered *Meatmaster* breeders plus commercial breeders
- ARC Performance testing of stud animals

### RESEARCH

Phenotypic and Genotypic typing of 4 Meatmaster sheep populations

- Karreekloof Meyerton Gauteng
- Bethulie Eastern Cape
- Hopetown Nothern Cape
- Prieska Northern Cape

## Progress

Representative blood samples from each Meatmaster sheep population were taken.

DNA Microsatellites are used to determine phylogenetic relationships and DNA profiles at the ARC Animal Genetics Laboratory at Irene.

Sheep blood samples  →  Extract DNA

Analysis  ←  Electrophoresis  ←  PCR

DNA Profiles

**Analyses of data is in progress.**

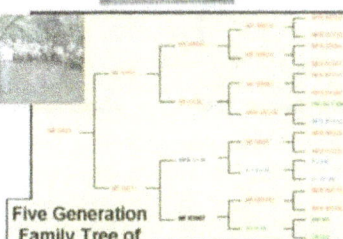

Five Generation Family Tree of a Meatmaster sheep

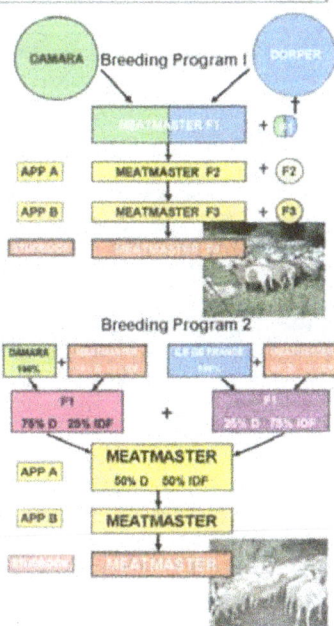

DAMARA   Breeding Program I   DORPER

MEATMASTER F1   +

APP A   MEATMASTER F2   + F2

APP B   MEATMASTER F3   + F3

STANDARD   MEATMASTER F4

Breeding Program 2

DAMARA 100%  +  ILE DE FRANCE 100%

F1 75% D  25% IDF  +  F1 25% D  75% IDF

APP A   MEATMASTER 50% D  50% IDF

APP B   MEATMASTER

STANDARD   MEATMASTER

### FUTURE

- Registration of more *Meatmaster* breeders with SA studbook and the ARC
- Registration of the *Meatmaster* breed by the National Dept of Agriculture
- National Sales of *Meatmaster* breeding sheep
- Export of *Meatmaster* sheep

CONTACT INFO:
- Tel: 051 821 1488; 051 654 0538; 016 365 5356
- Web: http://www.meatmaster.org

# Appendix D Presentation of Meatmaster research results - University of Johannesburg - Zoology symposium, 26 March 2009

**Meatmaster sheep breed establishment and development based on agro-economic principles, national standards, phenotypic and genetic characterization**

## OBJECTIVE

To establish the Meatmaster as a new locally developed meat sheep breed suitable to diverse ecosystems as production environments

## VISION

Durability and Fertility **+** Conformation

**=** **Meatmaster**

## HISTORY

| CR COLLETT | W VERMEULEN | C DU TOIT | FW PETERS |
|---|---|---|---|
| Damara+Dorper | Damara+Van Rooy+Dorper | Damara+SA Mutton Merino | Damara+Ile De France |

**PLUS OTHER INTERESTED BREEDERS**

**VARIETY OF BREED NAMES:**

*DAMARA CROSSBREDS*

*ILE DA MARA*

*BOSRANDER*

*MEATMASTER*

## RESULTS
### ACHIEVEMENTS

Meatmaster Sheep Breeders' Society of South Africa (Established 4 Feb 2005)
First Exco of Society (Elected 16 Feb 2005)
Breed Standards (Approved 31 May 2006)
First National Sale of Meatmasters (9 Aug 2006)
First Exports to Botswana (14 Jan 2007)
Breed Proclaimed (Gov Gazette 25 May 2007)
First National Championships (27 May 2008)
Certificated Meatmaster Courses (22 Aug 2008)
Meatmaster Breeders grow to 60 (26 Feb 2009)

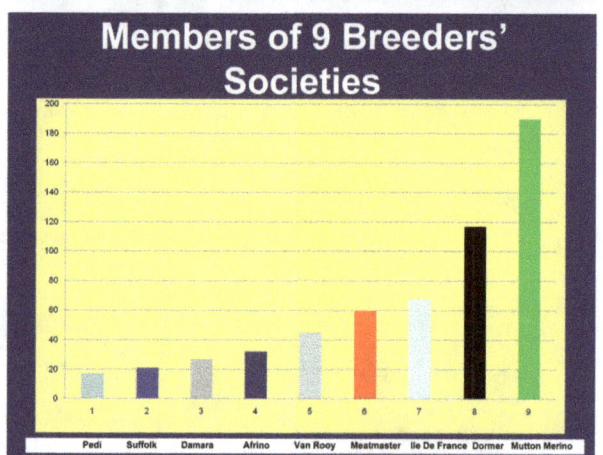

## Members of 9 Breeders' Societies

Pedi | Suffolk | Damara | Afrino | Van Rooy | Meatmaster | Ile De France | Dormer | Mutton Merino

# RESULTS
## Breed Establishment

The Meatmaster sheep is established as a new meat sheep breed ranking as the 5th largest Breeders' Society out of the 10 registered meat and dual purpose sheep breeds in South Africa.

**MEATMASTERS WILL TAKE SHEEP FARMING TO NEW HIGHTS**

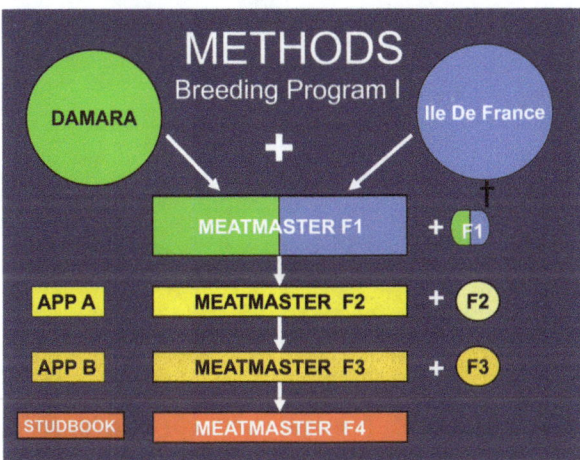

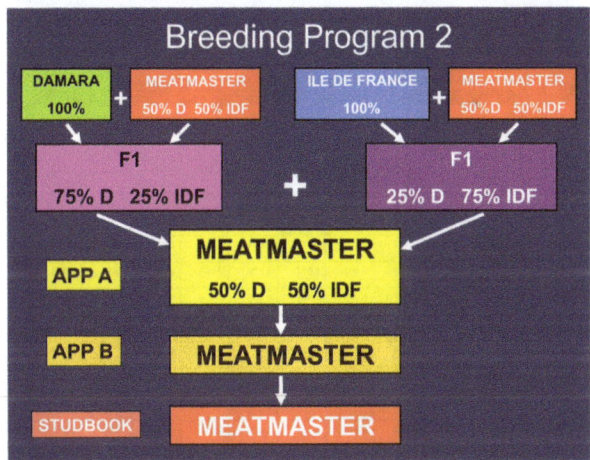

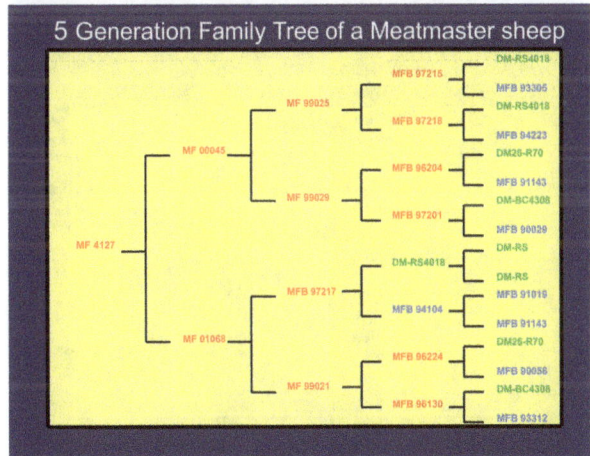

## METHODS
### Breeding Program I

DAMARA + Ile De France

MEATMASTER F1 + F1 †

APP A — MEATMASTER F2 + F2

APP B — MEATMASTER F3 + F3

STUDBOOK — MEATMASTER F4

## Breeding Program 2

| DAMARA 100% | + | MEATMASTER 50% D 50% IDF | | ILE DE FRANCE 100% | + | MEATMASTER 50%D 50%IDF |

F1 — 75% D 25% IDF + F1 — 25% D 75% IDF

APP A — MEATMASTER 50% D 50% IDF

APP B — MEATMASTER

STUDBOOK — MEATMASTER

## 5 Generation Family Tree of a Meatmaster sheep

MF 4127

MF 00045 — MF 99025 — MFB 97215 — DM-RS4018 / MFB 93305
MFB 97216 — DM-RS4018 / MFB 94223
MF 99029 — MFB 96204 — DM2S-R70 / MFB 91143
MFB 97201 — DM-BC4308 / MFB 90029

MF 01068 — MF 97217 — DM-RS4018 — DM-RS / MFB 91016
MFB 94104 — MFB 91143 / DM2S-R70
MF 99021 — MFB 06224 — MFB 90058 / DM-BC4308
MFB 96130 — MFB 93312

## GENOTYPING
### Meatmaster sheep populations

- *Kareekloof Meyerton Gauteng*
- *Bethulie Eastern Cape*
- *Hopetown Northern Cape*
- *Prieska Northern Cape*

## ARC GENETICS LABORATORY

**111**

## Genetic diversity (Alleles)

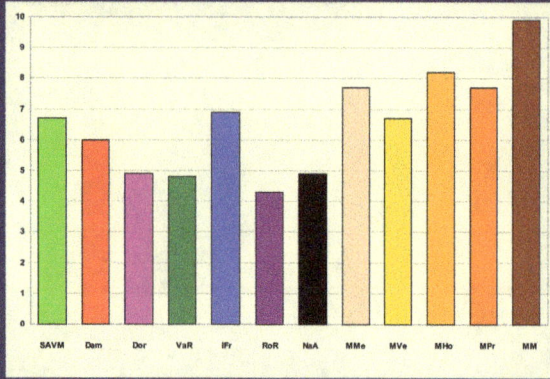

Table 3 Genetic differentiation between 11 sheep breeds and populations, with RST values below the diagonal and FST values above the diagonal. Values blocked in green indicate drift between specific Meatmaster populations and parent breeds; values blocked in blue show drift between Meatmaster populations and an outgroup.

|     | SAM | Dam | Dor | VaR | IFr | RoR | MMe | Mve | MHo | MPr | NaA |
|-----|-----|-----|-----|-----|-----|-----|-----|-----|-----|-----|-----|
| SAM | -   | 0.158 | 0.206 | 0.175 | 0.091 | 0.213 | 0.111 | 0.151 | 0.138 | 0.098 | 0.158 |
| Dam | 0.159 | -   | 0.133 | 0.104 | 0.167 | 0.190 | 0.047 | 0.066 | 0.069 | 0.065 | 0.132 |
| Dor | 0.180 | 0.119 | -   | 0.119 | 0.169 | 0.236 | 0.086 | 0.092 | 0.062 | 0.094 | 0.142 |
| Var | 0.163 | 0.035 | 0.083 | -   | 0.159 | 0.204 | 0.071 | 0.091 | 0.067 | 0.092 | 0.163 |
| IFr | 0.100 | 0.222 | 0.154 | 0.177 | -   | 0.242 | 0.083 | 0.133 | 0.110 | 0.096 | 0.151 |
| RoR | 0.127 | 0.177 | 0.293 | 0.203 | 0.242 | -   | 0.155 | 0.189 | 0.194 | 0.175 | 0.205 |
| MMe | 0.051 | 0.107 | 0.164 | 0.115 | 0.082 | 0.104 | -   | 0.026 | 0.033 | 0.026 | 0.100 |
| MVe | 0.135 | 0.128 | 0.175 | 0.120 | 0.108 | 0.164 | 0.028 | -   | 0.012 | 0.025 | 0.120 |
| Mho | 0.079 | 0.080 | 0.133 | 0.088 | 0.085 | 0.119 | 0.004 | 0.010 | -   | 0.021 | 0.093 |
| MPr | 0.088 | 0.118 | 0.194 | 0.116 | 0.098 | 0.102 | 0.017 | 0.016 | 0.009 | -   | 0.095 |
| NaA | 0.271 | 0.105 | 0.217 | 0.059 | 0.368 | 0.284 | 0.275 | 0.309 | 0.237 | 0.262 | -   |

Table 5 Proportion of membership of 11 breeds and populations of sheep to 10 nominal clusters, based on Bayesian analysis. Values in colored background indicate clusters dominated by established parent breeds, values in red background indicate three clusters of predominantly Meatmaster individuals.

| Cluster: | | | | | | | | | | |
|------|---|---|---|---|---|---|---|---|---|----|
| Pop: | 1 | 2 | 3 | 4 | 5 | 6 | 7 | 8 | 9 | 10 |
| SAM | 0.010 | 0.012 | 0.869 | 0.030 | 0.008 | 0.012 | 0.026 | 0.011 | 0.010 | 0.012 |
| Dam | 0.024 | 0.018 | 0.011 | 0.036 | 0.020 | 0.030 | 0.010 | 0.792 | 0.034 | 0.024 |
| Dor | 0.029 | 0.018 | 0.012 | 0.031 | 0.830 | 0.017 | 0.014 | 0.011 | 0.018 | 0.018 |
| VaR | 0.849 | 0.021 | 0.012 | 0.014 | 0.018 | 0.014 | 0.033 | 0.017 | 0.01 | 0.012 |
| IFr | 0.011 | 0.037 | 0.031 | 0.045 | 0.015 | 0.028 | 0.787 | 0.013 | 0.017 | 0.015 |
| RoR | 0.011 | 0.929 | 0.007 | 0.008 | 0.006 | 0.006 | 0.005 | 0.007 | 0.011 | 0.009 |
| Mme | 0.032 | 0.027 | 0.036 | 0.150 | 0.040 | 0.443 | 0.109 | 0.065 | 0.081 | 0.016 |
| MVE | 0.025 | 0.014 | 0.010 | 0.113 | 0.026 | 0.228 | 0.014 | 0.091 | 0.466 | 0.013 |
| Mho | 0.084 | 0.017 | 0.025 | 0.244 | 0.102 | 0.082 | 0.032 | 0.074 | 0.314 | 0.026 |
| MPr | 0.027 | 0.017 | 0.046 | 0.277 | 0.074 | 0.094 | 0.031 | 0.118 | 0.293 | 0.021 |
| NaA | 0.036 | 0.016 | 0.012 | 0.016 | 0.021 | 0.027 | 0.010 | 0.015 | 0.016 | 0.831 |

## BODY CONFORMATION

### RAMS
- Big and Muscular
- Especially strong over the loins.
- Excellent balance between length, width and depth

### EWES
- Sleek and Feminine
- Well formed udder
- Excellent mothering ability.

## IMPORTANT ATTRIBUTES TO NOTE

## NON SELECTIVE GRAZERS

- Excellent utilization of dry veldt
- None selective grazers
- Eats grass, bush, trees and shrubs

## Evaluation of MEATMASTERS

## Lamb % Fecundity

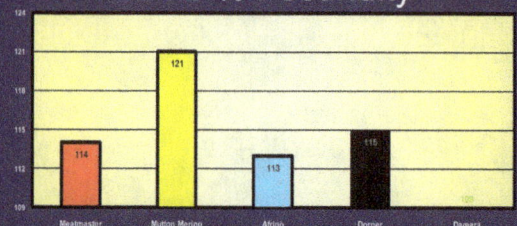

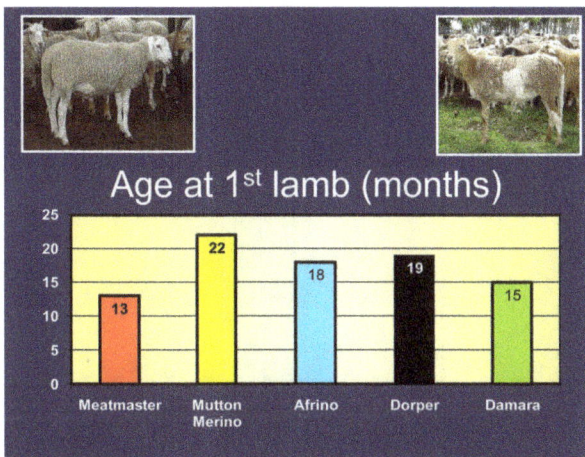
Age at 1st lamb (months)

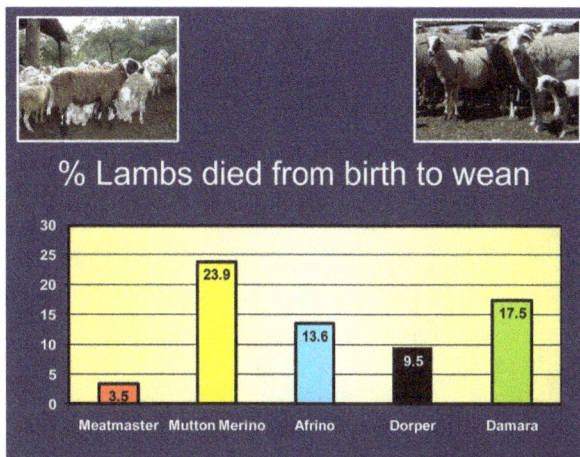
% Lambs died from birth to wean

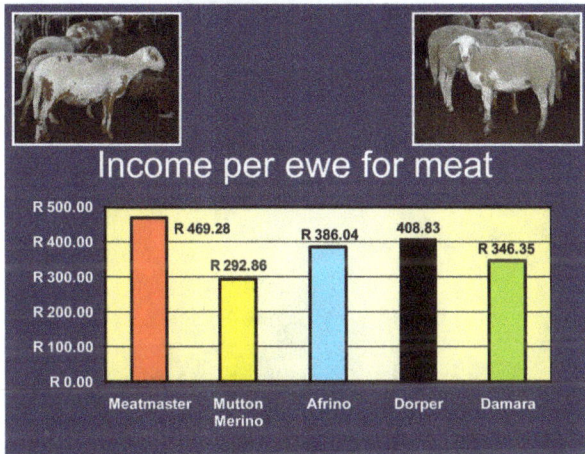
Income per ewe for meat

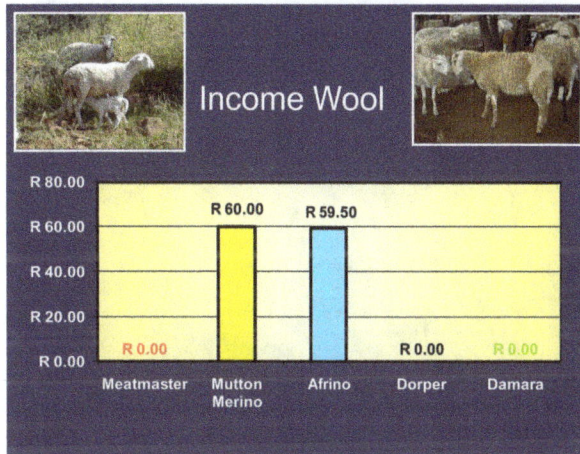
Income Wool

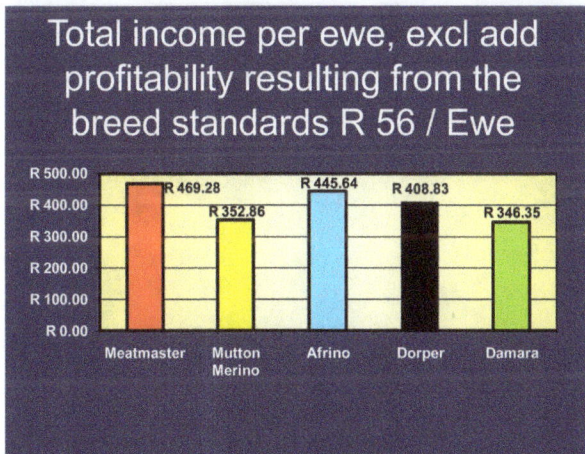
Total income per ewe, excl add profitability resulting from the breed standards R 56 / Ewe

MEATMASTER COMPARED TO DAMARA & ILE DE FRANCE

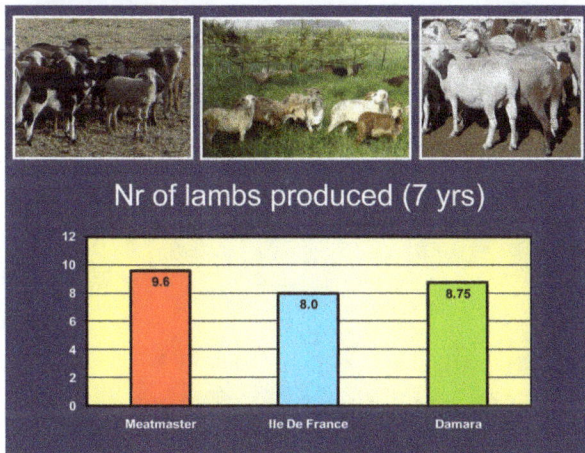
Nr of lambs produced (7 yrs)

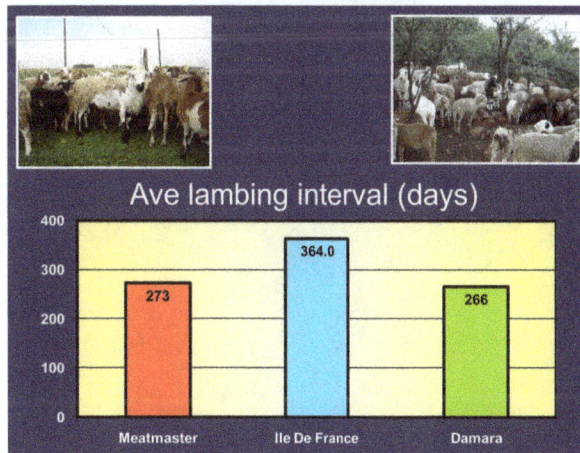
Ave lambing interval (days)

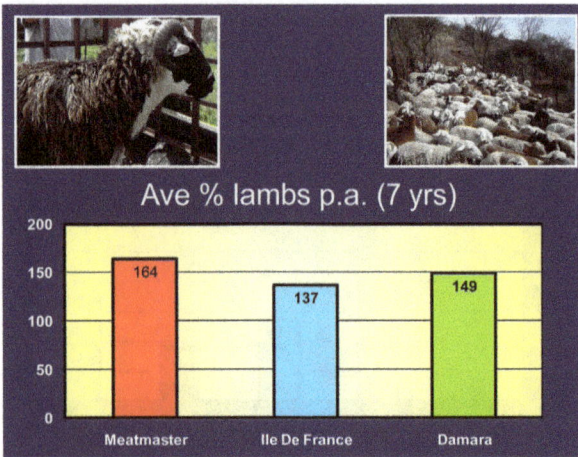

Ave % lambs p.a. (7 yrs)

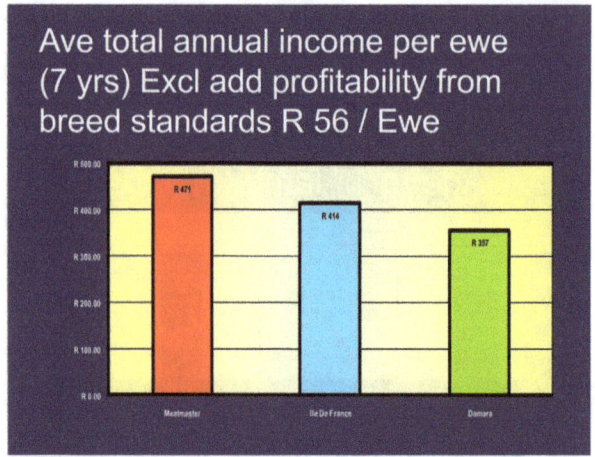

Ave total annual income per ewe (7 yrs) Excl add profitability from breed standards R 56 / Ewe

A New Composite Landrace Breed

MEATMASTER ewes with triplets

MEATMASTER 2002

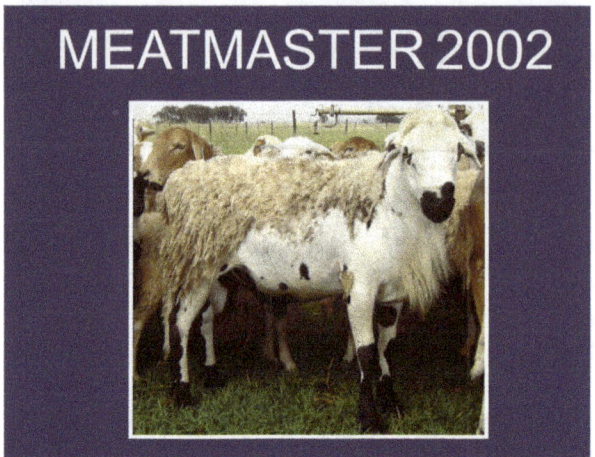

MEATMASTER 2005

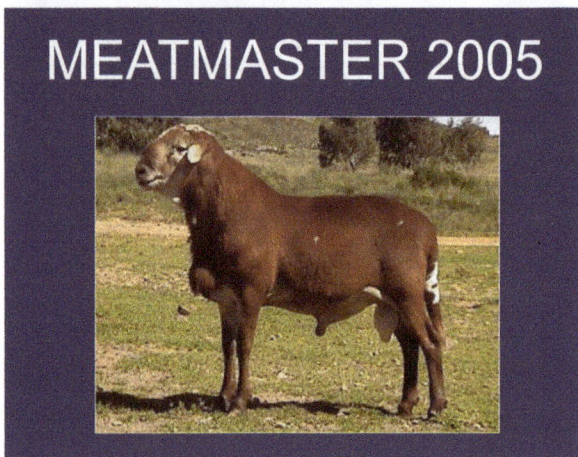

MEATMASTER 2008

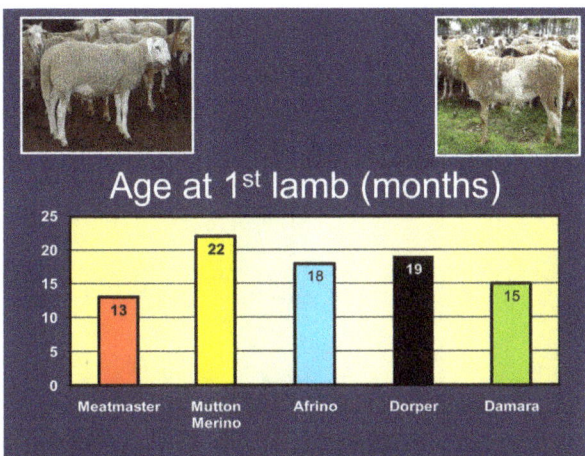

### Age at 1ˢᵗ lamb (months)

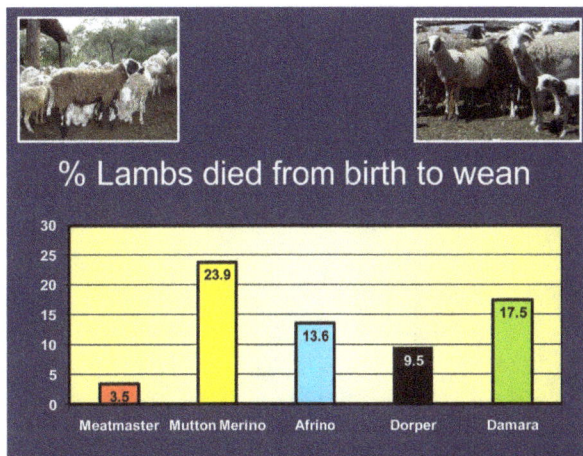

### % Lambs died from birth to wean

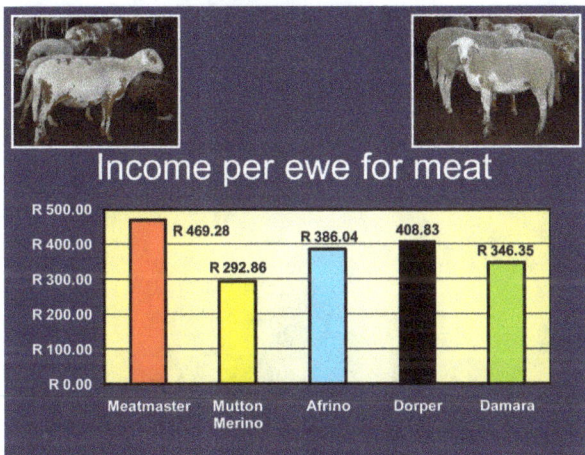

### Income per ewe for meat

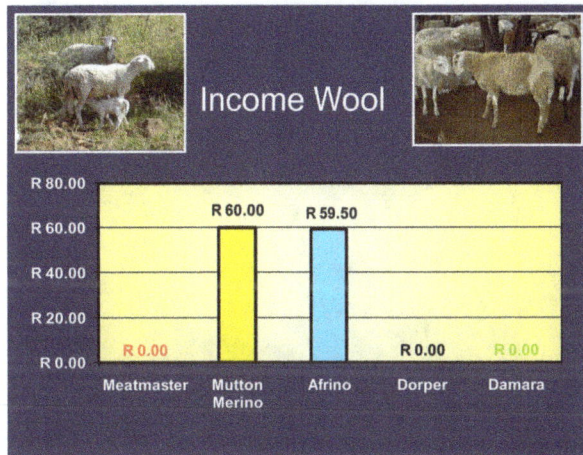

### Income Wool

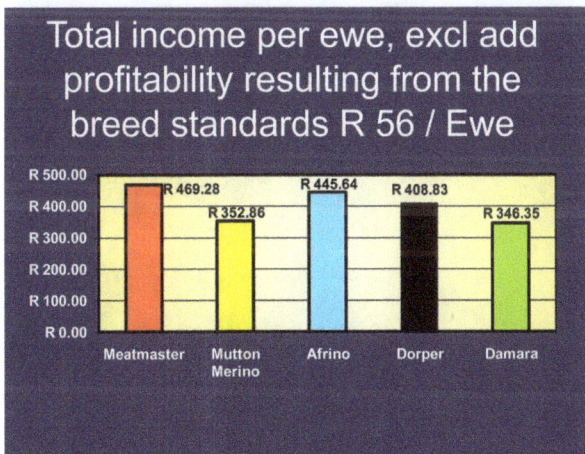

### Total income per ewe, excl add profitability resulting from the breed standards R 56 / Ewe

### MEATMASTER COMPARED TO DAMARA & ILE DE FRANCE

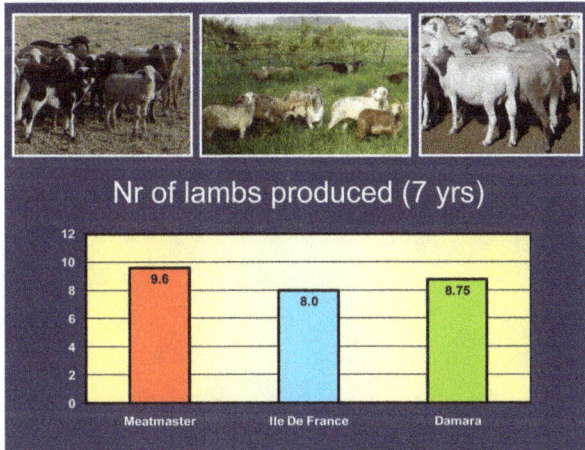

### Nr of lambs produced (7 yrs)

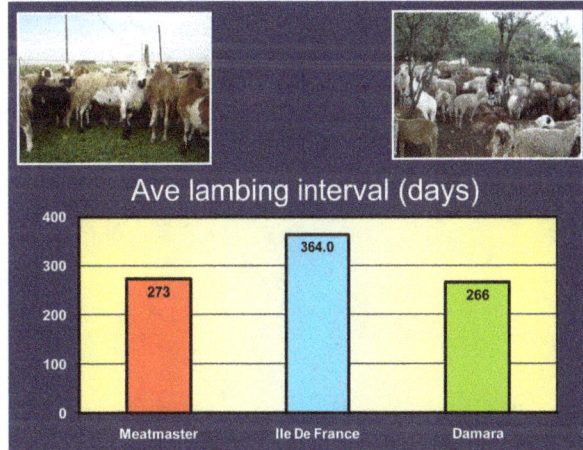

### Ave lambing interval (days)

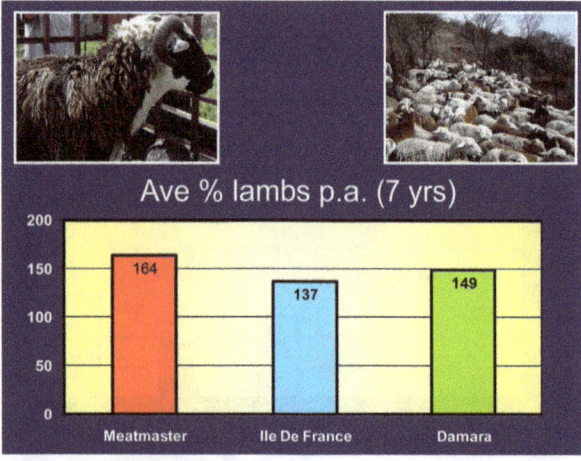

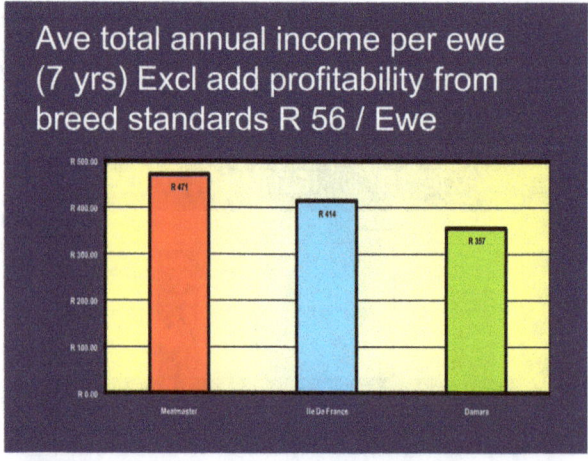

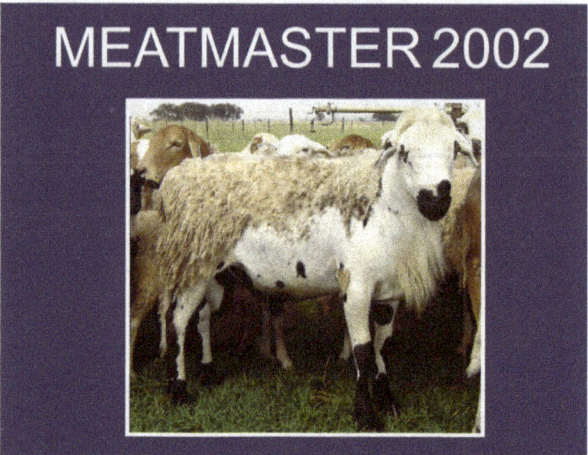

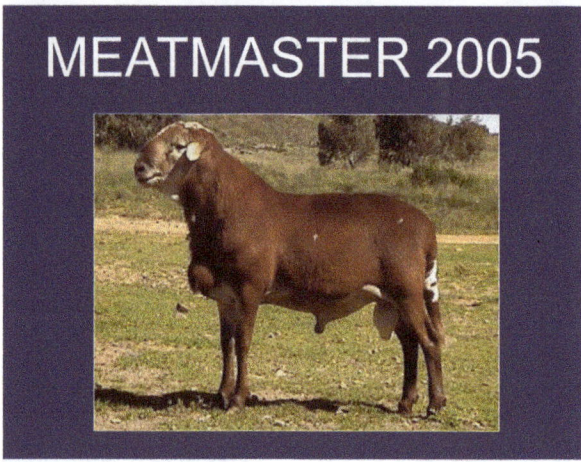

# Appendix E   Landbouweekblad article 29 September 2000 - J Cilliers

'n Skaapboer wat gereeld onder veediewe deurloop, beraam gou planne. Hierdie boer het besluit om 'n nuwe ras te teel wat in die bosse en rante kan oorleef en nie maklik in die veld gevang en opgelaai sal kan word nie.

■ Deur JAN CILLIERS

## Potslagters
## sal teëspoed kry met
## ranteskape

BO: Mnr. Peters probeer 'n skaapras teel wat ewe tuis in die bosse en in die rante is en wat dus gehard is en op die veld kan oorleef. Hierdie troppie van hom raak feitlik weg in die veld en is hoofsaaklik blaar- en struikvreters.

LINKS: Die skape verskil in kleur nogal baie, soos dié ooie aandui.

'n Damaraskaap laat hom nie sommer maklik in die veld vang nie, maar 'n Ile de France kan aangekeer word. Dus, wanneer 'n mens in 'n distrik met Ile de France-skape boer waar dié diere gereeld van die veld af verdwyn, kan met nuwe oë na die Damara gekyk word.

Dit is presies wat mnr. Freddie Peters van Meinfred-Boerdery aan die voet van die Suikerbosrand in die Vereeniging-distrik gedoen het.

Volgens mnr. Peters het hy dit begin oorweeg om 'n skaap uit die Ile de France en die Damara, wat gehard is, te teel. die skaap moet in die veld kan oorleef en wild genoeg wees om van veediewe af weg te kom. Hy het die teeltgeskiedenis bestudeer van die Dorper, wat uit imheemse Swartkop Persie-ooie en Britse Dorsethornramme ('n vleisras), ontwikkel is.

Mnr. Peters sê dit het hom laat besluit om 'n nuwe ras te ontwikkel wat in die bosse

sowel as in die bergrante, waar hy boer, tuis en aangepas sal wees. Die naam wat hy gekies het vir die nuwe ras wat hy wil ontwikkel, is dus heel gepas, die Bosrander.

Vir sy teelprogram gebruik hy vyf uiteenlopende, inheems aangepaste Ile de France-ooibloedlyne en drie onverwante Damararambloedlyne.

Volgens mnr. Peters is hy nog nie gereed om die Bosrander as nuwe ras te registreer nie, maar hy is heeltemal verras met die re-

44

REGS: Tweelinglammers by ooie op die veld kom algemeen voor, soos by hierdie ooi met haar twee lammers. Die ooie lam vir die eerste keer van ouderdom 13 maande tot 20 maande op die veld.

ONDER: Hierdie ram van 21 maande oud trek die skaal al op 'n stewige 61 kg.

sultate wat reeds behaal is. Die skape wat hy tot dusver geteel het, sal uiters geskik wees vir toestande oral in Afrika, is sy mening. Hoë vrugbaarheid is baie duidelik en moedereienskappe is by die ooie sterk gevestig.

Die skape het 'n bedekking van 50 % wol en 50 % haar, en dié vag word self afgeskud as die skape nie geskeer word nie. Die bene en kloue is sterk en die kleur wissel van spierwit tot swart en wit tot bruin. Die mees algemene kleur is 'n wit skaap met kleinerige bruin of swart kolletjies. Sommige skape het horings, ander nie, terwyl die sterte lank en smal is en nie 'n tipiese vetstert het nie. Goeie pigment kom voor en kom belletjies (onder die keel) is net by sommige teenwoordig.

Volgens mnr. Peters is die lammers klein by geboorte en weeg van 2,5 kg tot 4 kg. Lamprobleme kom glad nie voor

(Na bl. 46)

*(Van bl. 45)*

nie. Die lammers groei ook vinnig vir die eerste 42 dae en 'n groeitempo van 300 tot 400 gram per dag word gehandhaaf, maar daarna neem dit geleidelik af, omdat baie ooie in hierdie tydperk weer gedek word. Die tussenlamtyd van sy ooie wissel van 180 tot 220 dae. Die kuddegemiddeld staan tans op 250 dae. Tweelinggeboortes is volop en wissel van 30 % tot 40 %, wat neerkom op 'n jaarlikse aanwas van meer as 200 % in veldtoestande, sê mnr. Peters.

Op ouderdom 42 dae weeg die lammers van 13 kg tot 20 kg en op 100 dae van 22 kg tot 30 kg. Op 12 maande is die gemiddelde gewig van die lammers sowat 47 kg. Volwasse ooie trek die skaal op 62 kg; dié met suiplammers op 55 kg en dragtige ooie op 70 kg. Ramme weeg op 12 maande reeds sowat 52 kg en op 24 maande 63 kg. Volwasse ramme weeg tot 83 kg.

Volgens mnr. Peters is die nuwe skaap wat hy teel, hoofsaaklik 'n vleisskaap. Vetaanpakking kom nie voor nie, maar dit is egalig oor die karkas versprei.

Die skaap wat hy besig is om te teel, het reeds onderskeidende karaktereienskappe, sê mnr. Peters. Die kudde-instink is baie sterk ontwikkel en die skape wei gewoonlik in 'n trop naby mekaar en nie wydverspreid nie. Soms verdeel die trop tydens weiding in twee kleiner troppe en word ooie en hul lammers dikwels só van mekaar geskei. Die ooie en lammers verlaat egter nie die onderskeie troppe om na mekaar te soek nie, maar wag tot die twee troppe weer bymekaar kom. Mnr. Peters sê dit wil vir hom voorkom of die ooie dié eienskap uitbuit om lammers te speen. Hulle speen die lammers wanneer dié sowat 100 dae oud is.

'n Interesante eienskap wat hy waargeneem het, is dat die ooie soms hul pasgebore lammers in die veld wegsteek, sodat

Die skape wei gewoonlik in 'n trop, maar die trop verdeel gewoonlik deur die dag. Die skape bly egter by hul trop vir die res van die dag. Ooie gebruik skynbaar dié eienskap om hul lammers te speen.

hulle saam met die trop kan wei. Op pad terug soek hulle dan weer die lammers op. In een geval het hy ontdek dat 'n ooi haar lam, waarvan die been gebreek was, drie weke lank in die veld versteek het. Sy het snags saam met die trop gaan slaap, maar die lam bedags gaan voed totdat hy só herstel het dat hy hom weer by die trop kon aansluit.

Volgens mnr. Peters slaap sy skaapkudde in die veld, maar nooit in laagliggende gebied nie. Hulle soek altyd 'n heuwel, berg of krans uit waar hulle oornag. Dié gewoonte help om die trop teen ongediertes en veediewe te beskerm, sê mnr. Peters. In nat weer bly die skape bo-op 'n berg totdat droë weer intree. Hulle kan van twee tot drie dae sonder water klaarkom.

Hoewel die skape gras vreet, is hulle hoofsaaklik blaar- en struikvreters, sê mnr. Peters. Hulle loop baie en verkies struike soos katbos en die blare van doringbome. In die winter verkies hulle bos- en ranteveld en in die somer en lente katbosse. In die herfs vreet hulle graag blare wat afgeval het.

Sover het die skape wat hy geteel het, goeie weerstand teen algemene siektes in sy omgewing en word hulle glad nie ingeënt teen siektes nie. Bloednier en bloutong kom nie in veldtoestande voor nie. Die lammers is tog vatbaar vir lintwurms, gevolglik behandel hy sy skape een keer per jaar hiervoor, terwyl hy 'n opgletdip teen bosluise toedien.

Hy gee geen byvoeding aan sy skape nie, net 'n sout- en fosfaatlek, sê mnr. Peters.

BO: Die skape het nie werklik vetsterte nie, maar lang, smal sterte. Die skape se vet is nie gelokaliseer nie en egalig oor die karkas versprei.

LINKS: Mnr. Peters by die eerste groot ram wat hy geteel het uit kruisings tussen Ile de France-ooie en Damararamme. Die ram weeg 83 kg.

REGS: Hierdie poenskopram wat mnr. Peters vashou, is 14 maande oud en weeg 58 kg.

www.ingramcontent.com/pod-product-compliance
Lightning Source LLC
Chambersburg PA
CBHW082306210326
41598CB00028B/4460